HANS-JOACHIM ZILLM

The Human History Mistake

Figure Credits

© *Zillmer Archive. Except: The INFO Journal 3, Ed Conrad 5–7; from Mike Dash (1997) 13; Forbes Collection 14; Boston Athenaeum 16; Bernard Moestl 20; Steven A. Austin (1994) 24; Klaus Vogel 26; Ted Bryant 27; National Museum Copenhagen 28: Mark Lembersky (2000) 29–32, 35; Prof. Karl Dietrich Adam 33, 34; from T. C. Blair (2002) 36; from Edgar Dacqué (1930) 38; from Peter Kolosimo (1991) 45; from Denis Sauart (1955) 46, 60, 61; Lahr/Foley (2004) 47, 49; Harun Yahya 48, revised; Marcel Homet (1958) 51, 57, 58, 60, 66, 68; revised after Leroi-Gourhan (1968) 52; revised after Thorne/Macumber (1972) 53; NOAA NGDC (15. 11. 1999) 54, adapted by Zillmer; from Otto Muck (1968) 55, 56; from Werner Mueller (1982) 62; after Colin Renfrew (2004) 63, 64; adapted from "Science" (Vol. 307. p. 1729) 67; NOAA NGDC (15.2. 2005) 69, adapted by Zillmer.*

Picture Credits

© *Zillmer, except: Josef Bayer (1927) 2, adapted by Zillmer; Pachur (2000) 3 left; Paul C. Sereno (2003) 4 right; Steven A. Austin (1994) 5, Baker (2002) 5 small picture; Zillmer 6 left, U. Tarchiani (1998) 6 right; Dominic Oldershaw 7, adapted by Zillmer; schoolbook "Biologie heute S II" (1998) 12; United Press International 13; Harun Yahya 14; Pascual Jordan (1966) 15; Steven A. Austin (1994) 16; Holger Preuschoft 17 centre; C. F. Howell (1969) 17 right; Cecil N. Dougherty (1978) 19 left; P. A. Vening Meinesz (1943) 20, adapted by Zillmer; Volker Ritters 23 all individual photos, adapted by Zillmer; recreated after Junker/Scherer (1998) 24; Joe Taylor (1999) 25; Josef Bayer (1927) 28; picture detail from Dorling Publishing (1994) 29; recreated after J.-J. Hublin (2004) 30; recreated after T. D. Weaver (2003) 31; Cottevieille Giraudet 32; from "Harper's Weekly" (1869) 34; Luca u. Francesco Cavalli-Sforza (1994) 35; from O. Muck (1976) 36; P. A. Carling (2002) 37, adapted by Zillmer; from K. Buelow (1952) 38; after A. Jacobi (1931) 39; from C. H. Lindroth (1957) 40; Rakel (1894) 41; Harun Yahya 47.*

HANS-JOACHIM ZILLMER

The Human History Mistake

The Neanderthals and other inventions
of the Evolution and Earth Sciences

Suppressed Facts,
Forbidden Proof, Phony Dogmas

With 69 photos and 49 illustrations

Translated by Dr. Divleen Bhatia

Original German Edition:
"Die Evolutions-Lüge. Die Neandertaler und andere Fälschungen der
Menschheitsgeschichte."
Published 2005 in Munich.

Fourth hardcover edition 2008 published in Germany, Swiss and Austria.
Translated in foreign languages: Bulgarian (2006), Polish (2006), Czech (2007) and in
production French (2009).

© Copyright 2005 by Langen Müller in F.A. Herbig Publishing House,
München (Munich), Thomas-Wimmer-Ring 11, D-80539 München/Germany
Telephone +49 (0)89-290880, Fax +49 (0)89-29088144,

http://www.herbig.net and http://www.zillmer.com

More information about the topics in this book: www.zillmer.com
Cover Image: Evolutionary Series (above) from "Die Chronik der Menschheit" (Paturi, 1997)

Order this book online at www.trafford.com
or email orders@trafford.com

Most Trafford titles are also available at major online book retailers.

Note for Librarians: A cataloguing record for this book is available from Library
and Archives Canada at www.collectionscanada.ca/amicus/index-e.html

Printed in Victoria, BC, Canada.

ISBN: 978-1-4269-2352-4 (sc)

Library of Congress Control Number: 2009914170

*Our mission is to efficiently provide the world's finest, most comprehensive book publishing
service, enabling every author to experience success. To find out how to publish your book, your
way, and have it available worldwide, visit us online at www.trafford.com*

Trafford rev. 12/29/2009

 www.trafford.com

North America & international
toll-free: 1 888 232 4444 (USA & Canada)
phone: 250 383 6864 ♦ fax: 812 355 4082

Contents

Prologue

The bomb has fallen - a shock for the paleoanthropologists and evolutionists, seeing their scientific disciplines and the theory of Man's evolution in Europe shaken to their very foundations. Almost unnoticed by the public, a report spooked through the radio and television news: "Numerous Stone Age skulls in Germany are probably much younger than was previously claimed" (*dpa*, 16 Aug 2004, 17:59h). This was, in fact, a sensational piece of news, but one which I had already predicted and discussed in my 1998 book "Darwin's Mistake".

What set off the bomb was the new dating of some of the bones collected at the University of Frankfurt by Britain's Oxford University. The results demand that the picture of the anatomically modern man - at least in the period of 5,000-40,000 years ago be redrawn. Importantly, there are hardly any remaining significant human remains from the period of 30,000-40,000 years ago.

It is amusing that the Neanderthal of Hahnöfersand is only 7,500 years old rather than 36,300, and the skull of the "oldest Westphalian" from Paderborn-Sande should now be known as the "youngest Westphalian", given that he is only 250 years old, not 27,400 and dates from around 1750 A.D. The same goes for the bone fragments from the famous Vogelherd Cave, not 32,000 years as claimed but only 3,900-5,000 years old. For these and other finds, a *little mistake* of 20,000-30,000 years in dating has been made. Are the corresponding geological strata in which they were found just as young - which of necessity has to be the case?

The age of the modern Man, who allegedly arrived in Central Europe 35,000 years ago and displaced Neanderthal Man has just been dramatically reduced. There have been no further bones found for 17,000 and more years of his existence. The oldest immigrants now come from the central Klausen Cave in Bavaria and are 18,590 years old. Otherwise, all the bone remains of our ancestors are far younger than 10,000 years old. And does this mean that Neanderthal Man, who is now 30,000 years younger, will be classified as a modern man or is he still a Neanderthal?

It is not only the age of most objects, and certainly all those which have been re-examined, that has to be reduced but those finds previously classified as Neanderthal are now proven to be a brazen instance of false labelling. In 1999 two Neanderthal bones found in the Wildscheuer Cave were re-examined. The skull fragments, which were discovered in 1997, turned out to belong to... cave bears. There are only two other bone finds in Germany which bear witness to the existence of Neanderthal Man: That at Neanderthal itself and an upper thigh bone found in the Hohlenstein-Stadel Cave in the district Alb-Donau-Kreis, if we disregard a milk incisor found at the Klausennische. These Neanderthals too should be re-dated.

It is the efforts of individual scientists, who are owed a debt of thanks for seeking out the truth, that this swindle has come to light. Nevertheless, the anthropological community deliberately concealed the swindle for decades. Former employees told *Der Spiegel* news magazine (34/2004) that the star Professor of Anthropology, Reiner Protsch von Zieten, simply dreamt up the age determinations. Amongst themselves, the staff used the word "protsch" as a synonym for "bending into shape" (in fact, "inventing"). For decades, this was covered up by the personnel because this systematic is systematic: Only by presenting false facts in the guise of truth – in films, books and magazines – could the public truly be *sold* the dogma of Man's descent.

Because of the interplay with the media's "evolutionist mass indoctrination", reason has, in a way, been bewitched. The simplest contradictions, which are immediately obvious to a thinking person, or corruptions of the truth, are no longer recognised as such. The individual believes that certain things in his mind are real and he is convinced that this is absolutely logical and rational. He has no doubt as to the rightness of his belief.

"When a white-robed scientist ... makes some pronouncement for the general public, he may not be understood but at least he is certain to be believed ... they have the monopoly of the formula: 'It has been scientifically proved ... which appears to rule out all possibility of disagreement" (Standen, 1950).

In order to unmask this "evolutionary mass indoctrination", this book avoids impersonal scientific treatments of obscure subjects, riddled with jargon. On the contrary, it brings together a large quantity of factual material and empirical facts from all over the world proving that evolution theory is a pseudo-science of unproven hypotheses based on scientific falsifications. The empirical evidence presented in this book slices through evolution theory which smothers our society like a fog, in order to break down mental prejudice.

The evolution lie discussed in this book refers in particular to the deception of *macroevolution*, i.e. evolution above species level (Mayr, 1991, p. 319): An ape never turned into a human! On the other hand, microevolution takes place every day in nature and artificially during breeding. The rules of heredity with their associated variations of species described by Mendel, which underpin this microevolution, are today one of the cornerstones of experimental genetics. If one also takes climatic influences and spatial isolation into consideration, new varieties of existing animals or humans will be created, but none which represent an evolution as defined in the theories of Charles Darwin.

1 The Fountain of Youth

"Some aspects of experimental earth and human history writing ... arouse particular public interest. Once could cite them as prime examples of Zillmer's fountain of youth for earth and the life upon it. Referring to the presumed co–existence of dinosaur and Man, Zillmer significantly shortens the time horizon for the evolution of life in certain of its forms", wrote Prof. Dr. Bazon Brock (2001, p. 16). This co–existence, which contradicts evolution theory, is supported by firm evidence: The dinosaur expert, Paul C. Sereno recently found the fossilised bones of dinosaurs, aquatic dinosaurs and enormous primeval crocodiles in the same surface layer of the Sahara, together with a fossilised cow skull and fossilised human bones. The age which is said to separate dinosaurs from Man – the Tertiary – is a phantom age.

Clairvoyants in Arizona

During research in the U.S. state of Arizona, my attention was drawn to an unusual find reported in the "Arizona Daily Star" newspaper on 23 December 1925. In my book, *Kolumbus kam als Letzter* (*The Columbus-Mistake*), I published drawings of three artefacts from this collection (Zillmer, 2004, Photo 70). At that time, I was unable to get photographs.

The Silverbell Artefacts, named after the find site at Silver Bell Road, near Tucson, Arizona, are made of lead. These mysterious artefacts – which were found during several excavations since the original discovery in 1924 – were exhibited and described by the *University of Arizona* in Tucson in 1925.

Did the Silverbell Artefacts, which are made of lead and carry Latin and Hebrew inscriptions, come originally from Europe? All of the significant cultures of the Mediterranean region and India were working up lead long ago. In ancient Italy, lead was used in grand style for the construction of water pipes, drinking vessels and plates. "The oldest inscribed Iberian memorials are lead plates" (Haarmann, 1998, p. 420), and a lead plate with an inscription praising the deeds of the dead man, were found in a Viking grave. Other excavations unearthed medieval amulets made of wood and lead, with Latin letters and runes (Düwel, 2001, pp. 227–302).

An analysis of the Silverbell Artefacts lead, undertaken in Tucson on 24 August 1924, showed a lead content of 96.8% with small quantities of gold, silver, copper and zinc. It was established that the original molten lead was smelted from ore which is found in the southwest of USA. It therefore appears that the artefacts were produced locally and not imported from across the Atlantic.

The find comprises over thirty artefacts: swords and religious crosses made of lead, some of which carry drawings and inscriptions. In connection with our present theme, a long–necked sauropod (dinosaur) depicted on a sword, plays an interesting role. A report which appeared in the *New York Times* on 23 December 1925 put the finds in the spotlight of American public attention and sparked a controversy amongst the experts. Dean Byron Cummings, the leading archaeologist at the *University of Arizona* was one of several experts who vouched for the authenticity of the finds.

But their whereabouts were unknown. It appeared that they were no longer at the University. Finally, a tip was received that they might be at the *Museum of the Arizona Historical Society* in Tucson, but an initial phone call there brought no results.

During my later visit to the Museum, I was told that the artefacts were stored in the cellar of the Museum and viewing was out of the question. When I mentioned that I had come especially all the way from Germany just to see these finds, my luck changed. An old woman led my wife and me into the underground passageways.

There it stood before us: a wooden chest. A mysterious feeling came over us like a mystery as I opened the chest. The Silverbell Artefacts, over 30 in number and allegedly from the year 800, were neatly laid in specially cut wooden frames, contained in several removable wooden drawers. I was allowed to photograph more than half of the finds (see photos 1 and 2). I was not allowed to see the remaining drawers because I hadn't applied in advance to the Museum management for an appointment.

I hoped to find out more about the circumstances of the find at the museum library. In a bag, I did indeed find original photos from the excavations, which took place over five years up to 1928. I was able to obtain detailed excavation reports, excavation sketches, further photos and a description of the artefacts found from the unpublished report of Thomas W. Bent (1964) who was involved in the excavations.

One find in particular fascinated me: a dinosaur is depicted on one of the lead swords in a sensational way. If the artefacts are fake, the forgers must have been a bit stupid, because dinosaurs were reconstructed for the first time in the mid 19th Century. If older depictions show these primeval creatures, humans must at some time have seen living dinosaurs, or they had even older pictures of them – which cannot however exist, if our world view is correct. The depiction of a dinosaur on an artefact dating back to 800 A.D. reveals that the find is probably a forgery. Or is it?

If one observes the drawing on the sword, which might represent an Apatosaurus or Diplodocus who roamed formerly in the southwest of northern America, one is struck by the posture of a four-footed sauropod. In my German book *Dinosaurier Handbuch* ("Dinosaur's Compendium") which appeared in 2002, I took a controversial stand on the posture of this primeval creature, based on latest research findings. These indicate that contrary to previous opinion, sauropods held their heads horizontally and could only lift them a little, as the neck vertebrae would otherwise have wedged together (Zillmer, 2002, p. 89ff.). The tail too, as an extension of the spine, was held horizontally, either balanced in the air or used for swimming, since only a handful tail drags marks were found amongst the innumerable fossilised footprints.

Accordingly, up to a few years ago, every scientific work and every museum showed dinosaurs with their tails dragging behind them and their heads held high, often in kangaroo pose. Skeletons in this posture are currently being reconstructed in museums all around the world, wherever funds are available. If the Silverbell Artefacts dug up in 1924 are faked, the mere depiction of a dinosaur would be an inexcusable error, because before 1800 there had been no dinosaur reconstructions and, secondly, if an artist had created the artefacts shortly before their official discovery at the beginning of the 1920s, he would have depicted the dinosaur in the posture generally held to be correct and postulated in scientific works from this time: with upright neck and dragging tail. If the artefacts are forgeries, then the original artist was a clairvoyant because he gave a correct anatomical portrayal of the sauropod, such as only started to become accepted 70 years later. Or are the artefacts genuine after all? Did people over 1,200 years ago know what dinosaurs looked like? Were there even some still alive at the time?

Do the circumstances of the finds confirm their authenticity? The excavation photos show that the controversial artefacts were firmly embedded in a cement–like layer, know by geologists as "caliche". This geological layer made of calcium carbonate is present in large areas of the south–western USA, forming a kind of natural cement layer, which is is known as "desert cement". Stephen Williams, Professor for American Archaeology and Ethnology at *Harvard University* says in his book Fantastic Archaeology that the Silverbell Artefacts are fake but wonders at the same time how the alleged forgers could have embedded these artefacts so firmly in the caliche layer that the "impression" of an untouched find site could be created, given that the excavations were carried out officially by archaeologists from the *University of Arizona* (Williams, 2001, p. 242).

It was established that these special caliche formations stretch over wide areas along the Tucson Mountains and thus do not represent any "selective" occurrence, possibly artificially created with artefacts embedded at the same time (Bent, 1964, p. 321). Dean G.M. Butler of the *College of Mines and Engineering* at the *University of Arizona* also confirms that the caliche apparently hardens quite slowly and that "there is no chance that the artefacts could have been embedded into this formation after the arrival of the Americans to this area" (Bent, 1964, p. 323, cf. p. 177).

Although I am of a different opinion with regard to the hardening time, because calcium carbonate hardens relatively fast, the firm embedding of the Silverbell Artefacts in the caliche formation represents a proof of their authenticity. On the other hand, if the age of the caliche formations is relatively young in geological terms, at most 1,200 years old, the desert surface above it would be even younger. Did the desert form so recently there? Are the geological layers far younger than stated by the geologists? Is geological evidence of allegedly long periods of time simply a misinterpretation? Are the propagated long periods of earth's history a fiction? Were geological structures which

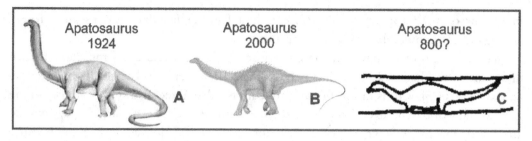

Fig. 1: Posture? *Anatomically correct, modern–looking representation (B) of a sauropod with straight neck and tail on a sword excavated in Tucson in 1924, allegedly dating back to the year 800 AD (C). According to the teachings prevailing in 1924, the sauropod would have had its head stretched up and a dragging tail (A) if the artefact were a fake.*

were supposedly created grain by grain, actually formed within a very short space of time by catastrophic events? Tsunamis, for example, change a landscape completely and create new geological layers within hours, whereby "normal" slow processes (such as sedimentation or erosion) would require millennia or even longer. Let us look at the supposed evidence that the earth's crust is very old, and then ask ourselves the question whether time–telescoping events might not serve to drastically reduce long geological periods. Because of science's obstinate linking of geological and biological (evolutionary) time ladders, the disseminated version of creepingly slow evolution is revealed as a fairytale, because analogous to the time–telescoping geological events, some sort of ape–ancestor must, so–to–speak, have mutated suddenly (as in time–telescoping) into modern Man.

The Phantom Tertiary

One argument that experts raise with regard to the rejuvenation of the earth's crust is the formation of huge fossil reefs. Were these geological formations really accrued by slow biological growth or was it a fast inorganic formation? The expert Julius Spriestersbach is of a different opinion with regard to the limestone (calcium carbonate) line of hills of the Rhenish Slate Mountains, hundreds of metres high and many kilometres in length, which the literature defines as fossil reefs: Spriestersbach says that in these unweathered "reef" limestone, "the joints between the layers look as if they were cut with a knife and the layer surfaces appear to have been smooth planed. This appearance runs counter to the theory of the natural growth of old reef formation" (Spriestersbach, 1942, p. 83). The only reef formation in the Rhenish Slate Mountains which Spriestersbach accept as a real coral reef lay at the end of the valley Aggertal Dam near Bredenbruch under water and for a long time could not be examined. Not until after the water was drained a 1985 examination shows that the supposed coral reef has no grown together colonies, but forcefully pressed successive layers" (Dr. Joachim Scheven in "Leben 4", 1992). Conclusion: there is no organic grown coral reef in the Rhenish Slate Mountains.

Similarly occurring Palaeozoic formations in central Sweden, England or the Alps are official interpreted as in situ grown (autochthonic) – a misinterpretation. It is said that the formation of those rifting requires millions of years to grow. But in these and other cases, we are talking about inorganically formed fibrous calcite (Stromatactis) which superficially look like a biologically (organically) grown reef structure, because the calcite layers penetrate the stone and are cross-linked in the cleavage fracture. In contrast to the organic coral reefs and bedded corral formation (Stromatoporoidea), the development of inorganic limestone (calcium carbonate) reef formations of these alleged coral reefs is very fast, as the water in the hollow parts of the stone can only be enclosed under catastrophic conditions (ibid. 1992). Conclusion: these ancient, supposedly autochthonous reefs were formed inorganically under catastrophic conditions by large masses of water. On the other hand, 400 million year old limestone from the dawn of the earth (Silurian and Devonian) does surely contain genuine corals, but these most certainly did not grow autochthonically. They are the more recent deposits of fossil stone from a "Deluge" which, naturally, contained pre–flood marine creatures.

One further argument in favour of an old earth crust: The partially still–extant reefs from the Tertiary supposedly indicate great age. During my visit to the *Great Barrier Reef* on Australia's east coast, I discovered that the age of this coral reef has been dated at 20 million years. "Impossible" in my view because the younger Tertiary was supposedly characterised by a climate which became increasingly cold; but warm water corals need a high average temperature of 20°C – a contradiction of the theory. To the astonishment of the experts, my opinion was essentially confirmed in 2001, when new analyses showed that the Great Barrier Reef is a youthful 600,000 years old – 33 times younger than was originally supposed (*Geology*, Vol. 29, No. 6, June 2001, pp. 483–486). However, the newly determined age of the reef, there were propagated the "Great Ice Age": It was even colder than it had been 20 million years before – too cold to grow as the coral.

Let's reduce the age of the reef still further and suppose that the corals grew in only a few thousand years before the Deluge, when the earth's axis was straight (perpendicular to the ecliptic) and that a global greenhouse climate existed from the North to the South Pole (discussed in detail in my book *Darwin's Mistake*) – in other words under climatic conditions as have been recently acknowledged to have existed for a period of time including the Cretaceous period (time of roaming dinosaurs) until to the middle of the Tertiary, 30 million years ago. But more about this later on.

Mountains too are becoming younger. Mica grains from the Pakistani Himalayan foreland have been dated to 40–36 million years (*Nature*, 8 Mar 2001, Vol. 410, pp. 194–197). The previous view was that the Himalayas folded 20 million years ago. In other words, the age is cut by one third.

For some time, new scientific examinations repeatedly indicate time reductions in the Tertiary (65 to 1.7 million years). In keeping with the arguments made in *Mistake Earth Science* (Zillmer, 2001), the Tertiary which followed the era of the dinosaurs should be reduced almost to "zero time" like a contracting rubber band. The effect which time–telescopes this Age (the Tertiary) is based on the apocalyptic scenario (Deluge), officially defined as the Cretaceous-Tertiary boundary (K-T boundary) at that point in time when the dinosaurs became extinct, 65 million years ago.

Major natural catastrophes always cause a leap in time (i.e. a time impact) for the areas affected, because cataclysmic events occur at a rapid time rate, effectively causing a speeding–up of the same geological sedimentation process occurring very slowly over a long period of time. If one fails to take this time impact into account, the natural catastrophe is a short–term representative for otherwise seemingly endless geological periods of time, which in turn have to serve as the basis for geological and biological developments, because evolution needs lots of time.

But there are thick stone layers which supposedly formed during the Tertiary. The layer arrangement (stratigraphy) for the Tertiary and succeeding Quaternary after the time when the dinosaurs became extinct (K-T boundary) is characterised correctly by the Professor of Geology, Kenneth J. Hsü: "Nowhere on earth could we find a continuous vertical sequence from today back to the age of the dinosaurs" (Hsü, 1990, p. 80). One must hereby bear in mind that the biggest sediment quantities of all time were formed during the Tertiary (after Holt, 1966).

The forerunner of modern geologists, Charles Lyell (1833, p. 15) already recognised that: The Tertiary formations were also found to consist very generally of detached and isolated masses, surrounded on all sides by primary and secondary (basement) rocks. Against these surrounding formations, the Tertiary formations place themselves as smaller or larger lakes and bays over the bearing them basement rocks. They are, like these waters, often very deeply, if at the same time "too limited in extent".

In this instance, Lyell's observations were correct, because Tertiary layers are not distributed over a wide area but lie scattered like the individual parts of a rag rug, as if one were scatter dominoes (= geological layers) with one heavy blow. In other words: in order to create a relative chronology of Tertiary layers, these would have to be aligned in some kind of logical way with one another, like stacked dominoes. But this prerequisite is entirely absent. If we look at geological maps of Europe, North America, South America or Asia and the distribution of marine deposits formed after the Cretaceous during the Tertiary, a system may be discerned, because a number of the biggest rivers drain broad or narrow basins, whose edges consist of concentric rings of bowl–shaped formations one inside the other.

A good example of this is the Paris Basin (Seine Basin), the edges of which were mainly formed during the age of the dinosaurs (Jura and Cretaceous) whilst the individual (younger) links of the Tertiary formations, covering increasingly narrow spaces, as concentrically shaped layers extending to the north coast, followed one after the other. The Thames Basin (England) is very similar. The Rhône and Danube flow out of long, extended Tertiary channels. In North America, the lower Mississippi Valley is covered with rings of Tertiary stone, as well as the east coast from Florida to Carolina. In South America, the Amazon flows through a broad basin of young marine deposits and in Asia the biggest rivers drain former sea floor. In summary: "It is hardly possible to define a greater coastal area of the present continents, which was not left, step by step, by the sea during the Tertiary" after the age of the dinosaurs (Walther, 1908, p. 455).

Analogous to the arguments in this book, it is on the one hand easy to recognise the flooding and later draining away of vast masses of water and, on the other hand, to recognise the formation or folding of the mountains after the dinosaurs died out (= end of the Cretaceous: K-T boundary), whereby the big rivers could only have formed at a time when there were no dinosaurs.

After the age of the dinosaurs, the earth's crust was moved on a wide scale and folded during the Tertiary. In Europe, the Alps, the Carpathians, the Apennines, the Pyrenees and numerous smaller mountain chains were formed. In Asia, massive mountain ranges were thrown up, whose folds pushed forward from the interior of Asia right through to the Indian and Pacific Oceans, like the waves of the sea.

In western North and South America, the great mountain ranges were formed. Through this process, the folding of the Andes, the direction in which the Amazon flowed was completely turned around; before then, this river rose from the Sahara and flowed not into the Atlantic as it does now, but across Africa and the still-connected South America, into the Pacific. This hypothesis which I introduced in *Mistake Earth Science* (Zillmer, 2001, p. 74 ff.) is based on finds of unweathered seashells and coastlines high up in the Andes and on traditional lore of the indigenous people, who obviously bore witness to the folding of the Andes. This hypothesis was underpinned by the investigations of the geoscientist, Gero Hilmer (*University of Hamburg,* Germany). His research was shown on German television under the title *The Ur–Amazon* ("Der Uramazonas", *Zweites Deutsches Fernsehen ZDF,* 24 Sep 2000, 19.30h).

On the surface of today's Sahara desert, the team of scientists found (still preserved) fossilised skeletons of aquatic Mosasaurs, which supposedly became extinct 130 million years ago, on the one hand and, on the other, live *desert* crocodiles in the remaining small lakes surrounded by the desert sand, which the Bedouins use as water reservoirs (gueltas). For how long can small populations of these animals survive in these tiny gueltas? Surely not 65 or 130 million years which is how old the Mosasaurs are supposed to be. How old is the Sahara? If swimming dinosaurs were still alive in the oceans a few thousand years ago – and maybe still are today – there's no problem: Mosasaurs became stranded in the forming desert sand and a few crocodiles survived in the small waterholes. Did the Sahara form only a few thousand years ago? Did a time impact occur here?

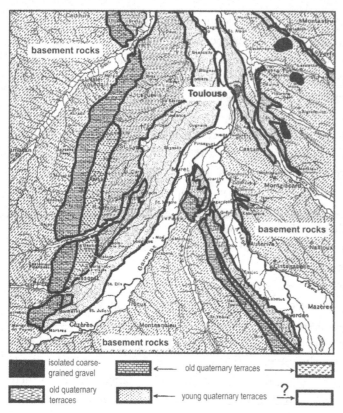

Fig. 2: Garonne Terraces near Toulouse. *Newer, thinner strips from the Tertiary and Quaternary form the edges of the big European rivers along the valleys in older basement rocks. It was always controversial as to whether the Garonne plain (light area) belongs to the newer Quaternary (Diluvium, cf. Bayer, 1927, p.88) or is even younger: Holocene (= Alluvium, cf. Obermaier, 1906, cited in Bayer, ibid.). Each terrace is said to represent a separate Ice Age. In fact, the terraces are the result of superfloods, shooting in through the valleys to the sea, which took place over short periods of time. The basement rocks remained above water level mostly.*

The transformation of the Sahara from subtropical steppe with hippopotamus, crocodile and elephant to a hostile sand desert, took place only 5,000 to, at most, 6,000 years ago (by geological timescale) as shown by the reconstruction of climatic conditions carried out in 1998 by the Potsdam Institute for Climate Research. The abrupt climate changes are said to have been caused by periodic fluctuations of the earth's orbit and the inclination of the earth's axis …

6,000 years ago, Lake Chad ("Chad" means "lake") still covered an area of 330,000 square km in the Sahara region. Today, mountains of marine chalk, so-called yardangs, rise out of the desert sand like icebergs floating in the sea, as witnesses to the biggest freshwater lake in the arid north of Sudan which still existed 5,000 years ago. Prof. Dr. Hans-Joachim Pachur of the *Freie Universität Berlin* confirms that there was an abrupt change of climate in the eastern Sahara: "The radiometric age of the organically bound carbon of the marine sediments ranges between 5,200 and 8,100 years (Pachur, 2002, p. 86). The geo-ecological reconstruction of the environment in today's hostile region is based, inter alia on finds of cattle bones, approximately 5,000 years old and similarly old ceramics. These are also confirmed by petroglyphs (rock carvings) found in the

Fig. 3: Climate Change.
Mountains of marine chalk (yardangs) rise out of the desert sand like icebergs. They bear witness to the freshwater lake in the eastern Sahara, which existed 5,000 years ago – in today hyperarid northern Sudan, Left photo from Pachur, 2000, p. 84.

surrounding mountains which show a woman milking a cow in the middle of a resting herd and "ceramic jugs are hanging from trees or posts, as mentioned by Bath 200 years ago in his description of a cattle–herding people 800 km south of Lake Chad" (Pachur, 2002, p. 86). It would appear that no cultural development has taken place since the formation of the Sahara. In any event, people experienced the formation of the Sahara. Large parts of today's Sahara still formed part of a large sea until relatively recently. On the other hand, Lake Chad was the source of the Amazon. At this time, Africa and South America were still one landmass. But at what point in time did these continents really break apart? Did all this happen only a few thousand years ago?

A dinosaur mass grave in Niger extends over a distance of 175 km. Today, the bones still rise out of the desert sand. In this state located west of Chad, one sees the scubby remains of the once extensive areas of water. Stranded here countless dinosaurs in the area of an ancient sea? Lake Chad is shrinking steadily since the formation of the Sahara. Since 1963 only four percent (now 1,000 square km) of 25,000 square km remain. In the last 6,000 years the lake's surface has become 3,300 times smaller. In other words, 6,000 years ago, the lake covered 3,300 times more land area. That's why dinosaur bones (including those of Suchomimus) are bleaching in the desert sands alongside the skeletons of huge crocodiles, such as Sarcosuchus, an alleged ancestor of today's crocodile which reached lengths of up to 15 metres. But here's the key point: You can't have crocodiles without water. Did these Cretaceous primeval crocodiles die out when the Sahara was formed, at most 6,000 years ago? The world-famous dinosaur researcher, Professor Paul C. Sereno of the University of Chicago, also found a cow's skull in the same geological surface layer – fossilised, just like the primeval crocodiles and dinosaurs in the same layer. Therefore, because of the fossilisation, it can not be of a recently deceased cow. Sereno himself wrote on his own homepage (www.paulsereno.org) in 2003: "What is a cow's skull doing in a place like this – in an area where one finds huge numbers of dinosaurs more than 100 million years old?" A refreshing question!

Where one finds a fossilised cow skull, fossilised human remains can't be far away.

Fig. 4: Surface Finds. *Huge skeletons from primeval crocodiles and dinosaurs rot in the Sahara Desert (left photo: Niger, Africa). The Sereno Team found a fossilised cow and fossilised human remains in these geological surface layers (right photo: Sereno, 2003).*

Accordingly, Sereno's team found fossilised human remains – in the same geological layer as the dinosaurs, primeval crocodiles and the fossilised cow. Were cow and humans mummified and fossilised by the burning heat caused by the desert formation, at the edge of the last remaining waterholes, while at the same time, crocodiles and dinosaurs became stranded in the drying lakes, where their bones still lie, well preserved, in the desert sand?

It appears that humans witnessed enormous cataclysms and transformations of the earth's surface. All along the East African Rift Valley system, whose length corresponds to one–sixth of the earth's circumference, the indigenous people also have an oral tradition which tells of great changes in the structure of the land and the formation of the Rift. This view is emphasised by geological phenomena: Some of the "Rift slopes are so bare and sharp that they must be quite young ... created during the human epochs" (Gregory, 1920).

J. W. Gregory (1894), the explorer of the East African Rift System, concurred in 1920 with the opinion of the prominent geologist, Professor Eduard Suess (1885/1909), which said that the developement of the East African Rift System was connected to the last great upheaval of the folding mountains in Europe, Asia and on the continent of America. Were the great mountains formed only relatively recently?

R.F. Flint (1947, p. 523) found that gigantic reshapings of the earth's crust took place during the age of modern Man: The earth was under pressure and its crust burst along a meridian almost the entire length of Africa ... The mountain range on the floor of the Atlantic could well have been formed at the same time; and the time of the crack formation and the folding must have been concurrent with the mountain–forming period in Europe and Asia. These mountains reached their present heights during the age of Man; the East African Rift System … was also largely formed during the age of Man, at the end of the Ice Age.

Tilt of the Earth's Axis

In my book *Darwin's Mistake* was pointed out that an abrupt change in the tilt of the earth's axis (with respect to the ecliptic) by around 20° was the trigger for the serious agitation of the earth's crust. In this case, the mountain ranges formed, so to speak, at breakneck speed: Humans saw the formation of the mountains (orogenesis). Moreover, the Polar zones iced over and the frost line suddenly shifted by several thousand kilometres. One of the effects of this was that the northern regions of Central Europe, as far as western Siberia, were shock-frozen.

For the first time brought the current investigations of geoscientists William W. Sager and Koppers Anthony P. a new scientific platform for discussion (*Science*, Vol. 287, 21 Jan 2000, pp. 455–459). They found that 86–82 million years ago, the earth possessed two magnetic Poles, 16–21 degrees apart from each other and they confirmed rapid polar wander episode … (implying) that the event was a rapid shift of the spin axis relative to the mantle (True Polar Wander), which may have been related to global changes in plate motion (plate tectonic, HJZ), large igneous province eruptions, and a shift in magnetic field polarity state (Sager/Koppers, 2000).

Did the dinosaurs survive a major tilting of the earth's axis – the earth's worst-case scenario – only to be wiped out as the result of a single asteroid impact? Surely not! If we place both of these events in a narrower period of time, 65 million years ago by geologic time scale, then these events initiated the end of the cretaceous period (K-T boundary) and the extinction of the dinosaurs. Bearing in mind that the Tertiary which followed the Cretaceous is an illusory classification (= phantom period) this cataclysmic scenario moves forward towards the present, to a time when humans saw the fast formation of mountains with their own eyes and kept their experiences alive in their mythology.

The actual folding of the mountains fits into the phantom Tertiary, which according to prevailing teachings was around 30 million years ago. The uplift rate at which the Alps are rising today is said to be only 0.8 millimetres per year (*Lexicon of Geosciences*, Germany). In contrast to this vision of mountains growing at snail's pace, sudden tilting of the earth's axis would cause the mountains to grow quickly in a kind of time–telescoping. In this case – in contrast to the slow rising process – there would not be enough time for erosion to take place: The steep mountain slopes still look "fresh". That is why the earth's crust is crumbling in our times, because it was newly formed only relatively recently; otherwise the Alp valleys should be full of stone rubble.

It is therefore not surprising that (not only) England's coast is crumbling rapidly (eroding), as shown by new data from the *European Space Agency* (ESA) satellite (*BdW*, 10 Aug 2001). Aren't the steep cliffs in their present form all still relatively young, at most thousands of years old? Weren't huge areas of land torn away by great floods as evidenced by the rapidly crumbling cliffs?

This view has been confirmed. David Smith, Professor of Geography at *Coventry University* presented his findings at an expert conference in Glasgow, Scotland. He claims that Great Britain was separated from the European continent by giant waves, after the Ice Age a few thousand years ago. An island (Great Britain) remained (*BdW*, 14 Sept 2001). Are the steep cliffs the silent witnesses of these recent events?

The stormy North Sea is also from a geological point of view a very young basin. The Rhine flowed through this former land lying above ancient sea level, now the floor of the North Sea and the river estuary was close to Aberdeen in Scotland (Basin Research, 13, 2001, pp. 293–312). At this time the Thames was a tributary of the Rhine. As late as the Bronze Age, the North Sea was still a fertile steppe and was then flooded (detailed discussion in my book *The Columbus-Mistake*).

The *Hamburger Echo* newspaper of 15 September 1951 (quoted in Meier, 1999, p. 490), carried a report on some seemingly curious finds: "The expedition ship, Meta, was able to make …. finds of incalculable value close to the Island of Heligoland. Two megalithic tombs were discovered at a depth of 30 metres in a mudbank. House remains, funerary objects, ancient tools and other utensils from the later Stone Age and the Bronze Age were recovered". Conclusion: the North Sea basin was flooded after the megalithic period.

In today's western American deserts there occurred superfloods as well, at the time when mountains rose. At the Yavapai Point Museum in Grand Canyon National Park it is written that the indigenous Havasupai – who live in the Grand Canyon – believe that (extract): "(the God of Evil) covered the earth with a great flood … Finally the flood waters fell and mountain peaks emerged. Rivers were created; and one of them cut the great gushing fissure which became the Grand Canyon".

The myths tell of real events. This means that humans saw dramatic changes to the earth's crust and were contemporaries of dinosaurs a relatively short time ago: Therefore the geological time scale is pulling together like an elastic band, thus reducing the Tertiary to almost "zero time".

What was the basis for the sub-classification of the Tertiary into long, individual periods (Palaeocene, Eocene, Oligocene, Miocene and Pliocene)? The breakdown of the Tertiary is based on the evaluation of a single incidence of a number of marine molluscs (molluscs). The proportions of molluscs allotted to each individual period also played an important role. In the 19th Century, the Tertiary was divided into three (now five) periods, whereby according to Charles Lyell, the Eocene should contain 5%, the Miocene 17% and the Pliocene 35–95% of living species: The lower the proportion of mollusc species, the older the layer was supposed to be. As may be expected, it was "quickly discovered that the proposed percentage figures couldn't even be carried forward from France to England" (Walther, 1908, p. 454).

Using this method, it is assumed that the death and reproduction rates, in particular with regard to molluscs (mussels), are the same all over the world. Despite this imaginary idea that the Tertiary had some kind of system, this arbitrarily dictated principle of classification and duration has persisted right up to the present day. Why do we hear only of aquatic animals in Tertiary layers, if mammals were supposed to have developed during this period? Why don't the fossils of land animals play any role in stratigraphy? "The material ... was only known to and accessible by the specialists..." (Thenius, 1979, p. 4). Secrecy is a must, as Tertiary layers are individually isolated, like oases in the desert and are apparently formed by the action of water. Land–living mammals must therefore have drowned during the Tertiary.

Stephen Jay Gould was able to show that every attempt to postulate a continuous evolution of mammals must conflict with the empirically tested material available (Gould, 1998). The family tree of the horse is a prime example of evolution. Gould (1998, p. 97) states: All important ancestral lines of the odd-toed ungulate (to which the horse belongs) are the pitiful remains of earlier, more profuse successes. In other words: today's horses are losers amongst losers – therefore almost the worst example for evolutionary progress, whatever the term is supposed to mean.

Superfloods

An iron asteroid with a diameter of only one km develops on impact the equivalent in energy of 1.55 billion tons of TNT, and produces a jet of water over 800 metres high, which spreads out at a rate of 600 km/h and even after 65 km, still retains two-thirds of its original height. The asteroid which hit the earth at the end of the Cretaceous and allegedly led to the extinction of the dinosaurs, is said to have had a diameter of 10 km.

The biggest wave ever observed, which took place after a landslide in Alaska in 1958, was around 500 metres high. In 1998, a 10–metre tsunami caused by an underwater landslide 2,000 km away was sufficient to kill more than 2,100 people in Papua New Guinea.

It is said that 200 million years ago, a meteorite impact triggered a tsunami with waves of up to 1,200 metres high which, in Central Europe, penetrated hundreds of km into the interior. The tsunami created stone deposits of the same age from Northern Ireland over South Wales into south–western England with a depth of up to 2.5 metres and even left layers of up to 20–30 centimetres deep in Pfrondorf, Germany. "But probably the meteorite was not solely to blame for the mass extinction", says the Tübingen-based geologist, Dr. Michael Montenari, "rather it was a combination of powerful volcanic activity and cosmic impacts" (*Spiegel Online*, 16 Sep 2004). This is a realistic description of an enormous natural catastrophe, but is it correctly dated as well?

Only 20 million years later, fish dinosaurs, marine crocodiles, spiny shark and long–necked dinosaurs (Plesiosaurs) were found piled up three–dimensionally – like sardines in a can – in a mass grave near Eislingen (Baden–Württemberg). Geoscientists at the *University of Tübingen* suspect that the release of methane hydrate was what triggered the ecological disaster. The release of methane hydrate is still underestimated causes of those catastrophes which have often take place during the earth's past. Untold quantities of methane gas lie beneath the North Sea. If this gas rises through the floor of the sea, huge masses of the sea floor – possibly as large as the size of Iceland – will collapse into the deeps like a landslide. This would cause gigantic killer waves. Within only a few minutes, tsunamis would crash over England, Holland and northern Germany. Cities such as Hamburg or Bremen would be destroyed.

The geologist, David Smith from the University of Coventry is convinced that such waves destroyed the land bridge between Great Britain and the Continent only 8,000 years ago. Happened at the same time that Iceland and the deep-sea floor of the North Sea, which now lies at a depth of 1,000–2,500 metres between the islands of Jan Mayen and Iceland, sank "by 2,000 metres in recent times" (Walther, 1908, p. 516)? Fridtjof Nansen proved in his expedition aboard *Fram* that are numerous ear stones (otoliths) which are derived exclusively from fish living in *shallow* seas were now to be found on *deep-sea* floor. But down there they decompose relatively quickly. Therefore, the sea floor sank in recent times dramatically.

Following a volcanic eruption on the island of Santorini some time around 1628, 60 metre–high tsunami waves probably flooded the eastern Mediterranean coasts and may also have wiped out Crete's Minoan culture. In comparison, the tsunami catastrophe of December 2004 in the Indian Ocean appears diminutive, even though it brought suffering to thousands.

If the Tertiary is an illusory geological classification (phantom age) and represents the time–impact of a natural catastrophe, these layers were mainly formed as the result of giant floods, as described above. One should also remember that side effects (earthquakes, volcanism and land upheavals) did not take place on a single day but were staggered over a longer period of time.

The natural concrete theory presented in *Darwin's Mistake*, proposed the rapid formation of sediment stone and geological layering caused by not only one or more major catastrophes, but also a series of subsequent cataclysms, which led to changes in the earth's crust and thus the development of "new", quickly formed, cement–like sediment layers. The idea that superfloods dramatically changed the face of the earth and created "new" layers containing "fossils cemented into them" (mass death), only a few thousand years ago, is only now for the first time supported by a new scientific study.

When, at the end of the (alleged) Ice Age, a 600 metre high ice dam, which contained

Fig. 5: Missoula Flood. *A few thousand years ago, after the "Ice Age", the Palouse Canyon in the US State of Washington was dug out in the "typical Ice Age" U-valley shape, from solid granite to a depth of 90-160 m, by the Missoula Flood hitting it with catastrophic force. Small photo (from Baker, 2002): this basalt block is 18 metres long and was transported 10 km by the Missoula Flood.*

the 270 km long Lake Missoula in today's US State of Idaho, broke, all of the seawater washed into the Pacific within two days. The torrential Missoula Flood contained more water than all the rivers in the world combined (*Science*, 29 Mar 2002, Vol. 295, pp. 2379–2380).

Victor R. Baker of the *University of Arizona* in Tucson confirmed in the scientific journal *Science* that geologists ignore the effects of superfloods because they assume that canyons and valleys were formed over millennia by the gradual effects of wind and water. For a long time, geoscientists were unable to believe that the entire landscape in the north-wester of the USA could have been completely transformed within only a few hours by a single event (*Science*, 29 Mar 2002, Vol. 295, pp. 2379–2380).

According to Baker, there were also superfloods on other continents, for example, Asia. The great basins with "Ice Age" lakes in Siberia (such as the Caspian and the Aral Seas) are witness to these floods, which even flooded highlands hundreds of metres high which stood as an obstacle in the pathway. The floodwaters cut channels into the mountain ranges, which can clearly be seen in satellite images of Central Asia. Baker mentions this extremely blinkered view and the thus very one–sided methodology of the geologists:

Methodological problems with the study of superfloods began early, at the inception of geology as a science. In the 1920s, Bretz's documentation of the spectacular effects of late-glacial flooding in the Channeled Scabland region in Washington State (*Journal of Geology*, Vol. 31/8, 1923, pp. 617–649), met with intense criticism from the scientific community.

Not until the 1960s was it generally accepted that this flooding was caused by catastrophic failure of the immense ice-dammed Glacial Lake Missoula along the southern margin of the Cordilleran Ice Sheet, which covered the north-western mountains of North America. These highly controversial studies of superfloods show that flood science has not achieved the universally accepted valid scientific, methodology envisioned by Charles Lyell. Instead, it is Baker´s view that superflood studies force us to look at the unexpected connections and explanatory surprises they engender (*Science*, 29 Mar 2002, Vol. 295, pp. 2379–2380).

Superfloods also caused the structural re-ordering of material torn away in the torrents which – once it had been washed into seas and oceans – formed new layers and thus sediment layers, which was hydrodynamically sorted and classified, based on the weight of the agitated and transported materials. The coarsest material (blocks and boulders) is at the bottom. Moving upwards, the grain sizes and thus any fossils contained in the sediment, become gradually smaller (gravels and sand).This system of layering repeats itself, so that depending on the number of giant flood waves which occur, several grain–size related layers lie stacked one on top of the other (sea figure 6).

In Australia too, there were superfloods. The geomorphologist, Professor Ted Bryant of the *University of Wollongong* in New South Wales and his colleagues suspect that the earth was hit quite regularly by large meteorites in the last few millennia. They came to this conclusion (which I share) following their investigation of multiple traces of destruction by giant waves (tsunamis) on the south-east coast of Australia (Bryant, 1997, 2001; cf. Young, 1996 and *Bulgarian Geophysical Journal*, 1995, Vol. XXI, No. 4, pp. 24–32).

On the basis of a computer simulation, Ted Bryant (2001) found that the height of the tsunami waves coincided with the impact of a celestial body with a diameter of 6 km in the middle of the Pacific. With the help of radiocarbon dating (which I reject), the age of the seashells which had been washed far inland was established: Te aquatic animals were washed on land in the last few millennia during the course of at least six different tsunamis. 6,000 years ago, and as recently as 400–500 years ago, the two highest waves even flooded areas which were 130 metres high (Bryant, 2001, cf. *Die Welt*, 20 Sept 2002).

Humans witnessed these giant floods and survived these cataclysms. This seemingly young age is probably still too high, because radiocarbon dating often results in fantasy figures. *Science* (Vol. 141, 16 Aug 1963, pp. 634–637) documented that a mollusc dated

Fig. 6: Sorting. When mineral settling sediment from mud, fine sand and gravel combines with organic swimming items and lots of water and is all stirred up together, after a period of settling, a vertical layering of the settling sediment according to grain size will take place (left). A corresponding hydrodynamic layering occurs after superfloods, such as the catastrophe in Versilia, Tuscany on 5 May 1988 (right): in this case, five sorted horizontal layers may be seen.

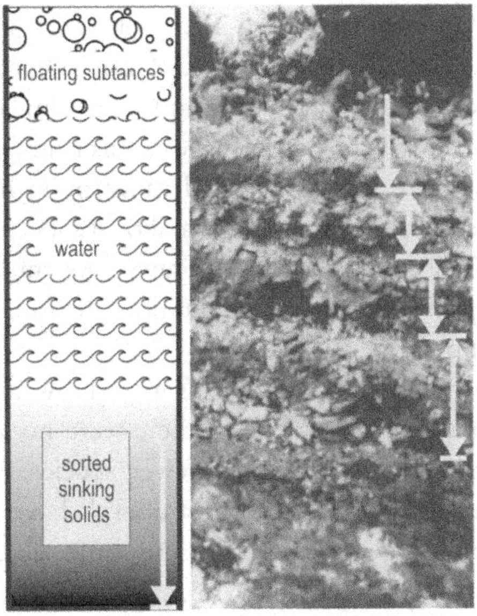

using the radiocarbon method was 2,300 years old. The only flaw in this was that the animal was still alive! In another case, testing established an age of 27,000 years. But in this case too, the specimen tested was still alive (*Science*, Vol. 224, 6 Apr 1984, pp. 58–61).

Further floods were proven, inter alia in the specialist journal *Geology* (Vol. 32, No. 9, September 2004, pp. 741–744): the island of Hawaii, the main island of the island group of the same name in the Pacific, was allegedly flooded 120,000 years ago by a gigantic wave. The water masses splashed the flanks of the Kohala Volcano, up to a height of 500 metres. The alleged cause of the giant wave was the collapse of the flank of the Mauna Loa Volcano.

Charles Darwin travelled around the world aboard *The Beagle* and on his way to the Galapagos Islands, visited the Andes in South America. In his travel journal he wrote:

The greater number, if not all, of these extinct quadrupeds lived at a late period, and were the contemporaries of most of the existing sea-shells. Since they lived, no very great change in the form of the land can have taken place. But what has exterminated so many species and whole genera? The mind at first is irresistibly hurried into the belief of some great catastrophe; but thus to destroy animals, both large and small, in Southern Patagonia, in Brazil, on the Cordillera of Peru, in North America up to Behring's Straits, we must shake the entire framework of the globe.

Based on his own observations, how could Darwin end up reaching such false conclusions? The accumulations of dead animals in South America, as elsewhere in the world, were known during Darwin's lifetime. Alfred Russel Wallace (1823–1913) who, like Darwin, similarly propagated the theory of natural selection, himself drew attention to the Siwalik Hills at the foot of the Himalayas, which were literally strewn with animal bones over a distance of many hundreds of kilometres.

One question which is often posed is: How could human beings escape this inferno? At this point, I would remind the reader that the Deluge did not happen in a single day.

Globally speaking, the various regions were affected by natural catastrophes and secondary events to differing degrees and at different times. The 2004 tsunami catastrophe in the Indian Ocean similarly showed that humans can miraculously survive, even where reason suggests this is impossible.

The idea of a global catastrophe horizon is backed up by a comprehensive genetic study: In the last few million years, mankind has almost become extinct at least once. This means that Man's ancestors must have lost a large part of their genetic variety – probably because the number of humans was drastically reduced (*PNAS*, 1999, Vol. 96, pp. 5077–5082).

The question remains as to how fast diversity of species can arise again after a major natural catastrophe. If there were no mountains or seas … and unlimited space, Nature would probably have fewer species. Where isolation continues for long enough, new species can form. But it doesn't have to take millennia: Two communities of a type of salmon, which lived in the same lake, were already going their separate ways after 60 years. In fewer than 13 generations two morphologically different populations developed from the descendants, who reproduced in almost complete isolation from one another (Hendry, A.P., et al in *Science*, Vol. 290, 20 Oct 2000, pp. 516–518).

On the Mediterranean island of Corsica there are two colonies of blue tits, which are separated from each other by only 25 km, but do not intermingle (SpW, 6 Aug 1999). The development of new species and thus the settlement of new habitats take place relatively fast, depending on various prerequisites (isolation, climate changes and environmental conditions).

In the second half of the 19th Century, biologists were astonished to see the genetic transformation over only a few decades of the light-coloured peppered moth into a dark brown variety, which had adapted to its soot and dust-soiled environment. Cases are not usually this simple, but brown bears and red foxes similarly developed white variants relatively fast in the Arctic. In keeping with the arguments presented so far, the Tertiary appears to be a period of time which has been shortened to almost "zero time" at all because of a time impact. This period ends with the start of the Quaternary 1.7 million years ago, and its sub-classification into Pleistocene and Holocene by 19th Century geologists was not based on any measure of time, but on events. The Pleistocene (now said to be 1,700,000–10,000 years ago) was called the Diluvium (*Latin*: flood), and the Holocene, which began 10,000 years ago was known as the Alluvium (which means "soil deposited by flowing waters"). This is the correct designation of the events; the geological layering as a result of superfloods occurred fast and not, as maintained by orthodox geology, grain by grain, over millions of years. The younger geological present since the end of the Pleistocene 10,000 years ago was also connected by earlier geologists with the floods and correctly defined, oriented on the events, because the Alluvium refers to all subsequent processes of floodwaters draining back into the oceans or local collecting basins, thereby filling the alleged "post–Ice Age" lakes like the Caspian Sea.

In the central Main Valley for example, around 50 metre–thick quantities of sand and gravel (Cromer Complex) were supposedly deposited extraordinarily quickly 85,000–76,000 years ago (Liedke, 1995) as mentioned by the Würzburg geologist, Erwin Rutte (1958): "The sediments were deposited in a single, uniform deposit phase without any significant interruptions" (Rutte, 1990, p. 235), "without interruption in a geologically very short space of time" (Körber, 1962, p. 30).

Huge layers, which also contained human relics such as the lower jaw of Homo heidelbergensis (Heidelberg Man) were formed in only a few hours or, at most, days, just beneath the modern surface of the earth (see photo 33). Taking Time Impact (fast deposit) into consideration, this event took place only a few thousand years ago.

Another example: "The bedrock of Berlin and the surrounding areas, which is close to the surface …. was formed, geologically speaking, only 10,000 years ago" (Bayer, 2002, pp. 29 and 35). The north–east German Basin is covered with a huge blanket of sand and clay which, partially layered, contains a massive quantity of larger and smaller blocks. In many cases these deposits show that they were transported, rolled and deposited there by raging floodwaters. For 19th Century geologists, these alluvial formations were connected with flood-like water movements (cf. Walther, 1908, p. 492).

Deep drillings have shown that the thicknesses of the "Ice Age" deposits in northern Germany are greater than those in most areas of North America (*American Geologist*, 1892, p. 296). The North German Plain (or Northern Lowland) is an older basin, up to ten km deep which was filled up with sediments and salt deposits. More recently, during the "Great Ice Age", sediment layers formed, which from Poland to Belgium, via Denmark, reached thicknesses of up to 200 metres, e.g. in Hamburg, up to 192.6 metres and in Berlin, 166 metres (Wahnschaffe, 1901, p. 17ff.). This cannot be debris left behind by glaciers, as the individual surface is mostly free of the relief forms of the deeper bedrock. Moreover, "crust movements have taken place very recently" (ibid. p. 70).

E. Boll's theory is interesting (1846, p. 263ff.). He claimed that in Scandinavia, as a result of volcanic activity, there was a catastrophic outcropping of granite, which he sees as the main reason for the Diluvium in northern Europe (cf. photos 39–42). He says that a rain of volcanic (stone) bombs, such as were also catapulted across the landscape during the Mount St. Helens eruption, poured down over the neighbouring countries. One can indeed find "erratic" (non-indigenous) blocks as far away as Thuringia, which come from Scandinavia and whose mode of transport is disputed. According to Charles Lyell, these blocks were transported to Germany from Scandinavia on icebergs but this drift theory, which is still supported by some geologists, has meanwhile been given up (Schwarzbach, 1993, p. 34).

One must underline the different views of geologists with regard to the deposition of geological layers: Up until almost the middle of the 19th Century, their formation was, on the one hand, said to be linked to specific events (flooding, volcanic eruptions, movements of the earth's crust, landslides) whilst on the other hand – since the introduction of geological time scale – the formation of sediment layers, assuming a (slow) rate of sedimentation, was interpreted as a gauge of the earth's age (indissolubly linked with evolution). Looked at in an abstract way, from this "modern" scientific viewpoint, the formation of geological layers is a function of long periods of time and not the result of events of short duration. These are two completely different viewpoints of the deposition and formation speed of the soil and stone layers, in particular since the extinction of the dinosaurs in the last 65 million years (Tertiary and Quaternary).

If one sees the stone layers from these periods of the earth's recent past as the result of one or more events (natural catastrophes), the corresponding geological layers must be subject to a time–impact because catastrophes, and floods in particular, characteristically cause fast and abrupt changes to the landscape through denudation and deposition. In recent years, however, a slow but steady shift in geological thinking towards (back to) the idea of Neocatastrophism, as postulated by Georges Cuvier (1769–1832) has been taking place. He shared the opinion of the geologists of the time – as do I – that animals of earlier times were made extinct by violent natural catastrophes (catastrophe theory) but also that the species cannot change (dogma of the *constancy of species*).

However, this slowly forming Neocatastrophism amongst geologists cannot on the one hand leave Charles Lyell's principles of uniformity as the basis of modern geology and thus geological-biological chronology untouched, while at the same time crying out like hermits in the desert: we believe in catastrophes. Globally speaking, this position contradicts the theoretical principles, in other words the dogmas of the theory of uniformity in geology and biology. If you're going to have a re-think – then at least be consistent! Let's take a closer look at a catastrophic event, i.e. a time–impact.

The famous chalk cliffs of southern England – similar to those on Germany's largest island Rügen (Rugia), and the Danish island of Moen – mainly comprise the remains of planktonic algae. There are also shells of various molluscs and of Foraminifers (single-cell animals with a shell). What is so unusual about this so-called writing chalk, is its high marine purity, as it doesn't contain any sediment from the mainland. That is why this limestone formation is puzzling from a modern geological viewpoint, especially since there is now nothing to compare with these extensive chalk deposits. The mass of chalky algae remains becomes absolutely incredible, when one works on the basis of creepingly slow processes. Was there, contrary to the dogmas of the theories of uniformity, an explosive blooming of algae?

Were these planktonic algae a source of food for the (in my opinion) aquatic sauropods? Is that perhaps the only reason why these herbivorous animals, which reached lengths of up to 50 metres, had long tails and necks, which moved only horizontally (not vertically), so that they could push these little creatures into their mouths, which contained a type of rake of peg-like teeth for filtering small animals, such as one sees today in whales?

The very high temperatures and the greenhouse climate prevailing during the

Cretaceous were the planktonic for the explosive blooming of algae, the global distribution of foraminifers and, ultimately, the sauropods. The tropical average temperatures in the central regions officially prevailed up to the middle of the Tertiary, 30 million years ago, before temperatures dropped. Thus, for example, plankton–foraminifers, which indicate sub-tropical temperatures, lived during the Eocene (earth age: 55 to 38 million years) in north-west Germany. To the surprise of researchers on an international Arctic drilling expedition, latest investigations show that 55 million years ago, sub–tropical sea algae swam not only in northern Germany but also in the Arctic Polar Sea in water temperatures of around 20°C (German *dpa* report on 6 Sep 2004, 14.11h) – cf. *Darwin's Mistake*, pp. 106 and 238f.).

In any event, only turbulent conditions during gigantic flooding are in the position to wash up massive chalk deposits. In the United States, isolated cross-layering in the chalk deposits has even been found. This confirms fast formation under water, with water speeds of 50–160 centimetres per second, depending on the depth of the water (*Sedimentary Geology*, 26, 1980; cf. Zillmer, 2001, p. 281). With higher water speeds, flat sand surfaces are formed; with lower speed, ripples are formed.

The massive chalk layers were washed up by the violent movement of water, whereby a time–impact by virtue of the fast formation has been documented, before enormous waves greedily dug into the chalk layers again like bulldozers, which is how these steep cliff lines were formed in the first place. It is therefore not surprising that (not only) England's coastline is rapidly crumbling (eroding) today, because these cliff lines are relatively young. This explains why the emblem of the Rügen chalk coast, the pinnacle "Wissower Klinken" whose point towers 20 metres high, collapsed. 50,000 cubic metres of chalk now lie on the beach and in the Baltic Sea.

Spillway Grand Canyon

Large masses of water are probably also responsible for the creation of other wonders of Nature. In *Darwin's Mistake* (1998, p. 223) I quoted the example of the Niagara Falls, which are apparently only a few thousand years old. At the same time I also stated that the Grand Canyon must also be relatively young (Zillmer, 1998, p. 229ff.).On the other hand, the theory of a slow development of the landscape through climatic erosion after the Cretaceous period suggests that the Grand Canyon is 65 million years old (time when the Kaibab Plateau lifted), but according to the more recent estimates of some geologists, the Canyon is only six million years old.

However, only very large, not small quantities of water (Colorado River) are responsible for the formation of the Grand Canyon. Accordingly this must be a relatively recently formed canyon, created catastrophically (fast) by the erosion activity of large masses of water in several phases – in other words – a time–telescoping time impact.

Fig. 7: Grand Canyon. *The rubble in the foreground is 54 metres above the Colorado River and is said to have formed only 160,000 years after the collapse of the lava dam (LD) (Fenton et al,. 2002, p. 196). Current opinion holds that this lava was thrown out during an volcanic eruption in the late Neozoic Period several million years ago (Hamblin, 1994). Given the good state of preservation of the almost unmarked rubble (B = basalt block approx. 1 metre in length), the author suggests that this event took place only a few thousand years ago.*

This opinion was exactly confirmed in 2002 by the US Geological Survey in conjunction with geologists from the *University of Utah*. *United Press International* reported this on 20 July 2002 and the fall-out followed in the *VdI News Germany* (*VdI Nachrichten*) on 4 October 2002 under the headline "Ancient Canyon? Wrong!"

Robert H. Webb of the *U.S. Geological Survey* in Tucson, Arizona reports that the lower third of the Grand Canyon, the "Inner Gorge" is probably only 770,000 years old: "The canyon was dug out by a series of short but intense events". Enormous lava dams are said to have blocked the western part on several occasions and retained the floodwater. When the dams then collapsed, gigantic flood waves dug out the river bed which is now the Grand Canyon.

The investigation paid particular attention to a flood wave which allegedly occurred 165,000 years ago and dug out part of the lower Grand Canyon. According to Webb, the water masses were 37 times bigger than the biggest Mississippi flood: more than 400,000 cubic metres per second (cf. Fenton et al, 2002, pp. 191–215). To get an idea of this: before the embankment dams were constructed, the Colorado River transported just under 6,000 cubic metres per second. Now it transports less than 400 – in other words one-thousandth of the quantity compared with at the time the Grand Canyon was formed.

The biggest flooding of the Grand Canyon is said to have taken place 120,000–100,000 years ago. It is interesting to note Robert H. Webb's finding, that because there is no river delta, larger lakes must have formed behind the lava dams only 11,000 years ago. I had already drawn attention to this fact, because the search for erosion mass which had been carried away remained fruitless. Downstream too, in the direction of California, there are indeed landfills, but the total quantity of material they contain is far too small. Was the eroded material washed by the floodwaters into the Pacific where it formed new layers, sorted by grain size?

The water of the giant waves which gushed through the Grand Canyon is said to be the melt water from the "Great Ice Age" ice sheet. Since, in my opinion, this period of almost 2 million years duration did not actually take place and the ice regions were much further north at the border with Canada, the only explanation – once again – is superfloods. It is interesting that Robert H. Webb draws comparisons with the Biblical Flood. However, in terms of the Flood, he sees only intense rainfall as the cause. But he is forgetting the "fountains of the deep", because the water also comes out from deep inside the earth's crust, perhaps even out of the earth's outer mantle.

To the amazement of geologists, lava volcanoes spew out large quantities of water and steam. According to current geophysical theory on how lava volcanoes function, there can be no expulsion of water. But there are also virtual mud volcanoes with and without lava expulsion, as was seen during the 1980 eruption of Mount St. Helens in the US State of Washington. During this event in 1980 pyroclastic streams surged down the side of the mountain with the speed of a hurricane as one flowing, turbulent mass of mud, made up of fine volcanic dust, once the ice and snow–covered mountain top had been blown off by an explosion of steam. One might expect that these deposits would be homogeneous and well–mixed. However, completely separate layers of coarse and fine particles formed from the fast-flowing mud. Such processes follow the laws of hydrodynamics.

On 19 March 1982, a large mass of snow which had accumulated in a crater, melted. In only nine hours, the mud flow dug out a system of canals and three canyons, 30 metres deep. One of these canyons was given the name "The Little Grand Canyon of the Toutle River", since it resembles the Grand Canyon on a scale of 1:40. Conclusion: a large quantity of water or mud can achieve extremely quickly what a small amount of water would need an eternity to do – a geological time–impact.

The soil also liquefies because of earthquakes. Images taken by the Landsat–7 satellite show how an earthquake in western India in January 2001 caused water to gush out of the ground, in places which were previously dry. Because of the violent agitation, localised liquefaction occurred releasing water from fine sediments (cf. *SpW*, 27 Apr 2001). Corresponding fossil earthquake springs with soil liquefaction (Thorson, 1986,

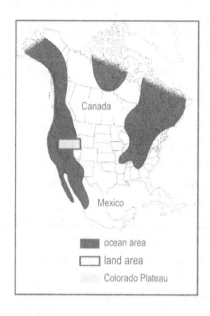

Fig. 8: Lifting. *The Colorado Plateau rose with the Rocky Mountains before the formation of the Grand Canyon.*

ocean area

land area

Colorado Plateau

p. 464f.) as found in the ice–free regions are, particularly in Germany, ascribed to the effects of the Ice Age rather than to earthquakes, and are known as ice wedges.

The thesis presented in this book concerning the "fountains of the deep" has meanwhile been confirmed by several scientific investigations. "There are probably vast water reserves 600 km below the earth's surface" (*BdW*, 27 Apr 1998) and the American geologist, Joseph Smyth of the *University of Colorado* at Boulder believes that the inside of the earth contains as much water as three to five oceans (*SpW*, 18 Sept 1999).

The "Drainage Basin-Theory" which explores this area in more detail was introduced in my book *Mistake Earth Science*. At temperatures of 425–450°C, the drainage basin causes the precipitation of minerals (iron, calcium, magnesium) in the region of the lower earth crust, whilst at the same time, steam pressure is built up because of the prevailing heat (Zillmer, 2001, p. 252ff.). Because of the escaping excess pressure sand masses are pushed to the surface with the water. The precipitated calcium in the drainage basin mixes with the water and sand to create a kind of cement, which because of the excess pressure can form a kind of dome on the earth's surface.

Because of the binding agent contained (calcium), this soft sand-calcium mixture hardens into a cement-like elevation, like Ayers Rock in Australia. Its smooth surface proves that the mountain was created "in one go" within a short space of time, since the teeth of time have not yet gnawed so deeply. The holes which are sometimes visible on the surface are essentially bubbles of water which dried out during the hardening process of the originally soft "natural cement mixture". An elevation such as Ayers Rock cannot be formed grain for grain, one reason alone being because of the enormous geological time-reducing hydraulic hardening process for calcium (lime).

Let's go back to the Grand Canyon. In the opinion of many geologists, the Grand Canyon region began to lift up around 65 million years ago, to a height of about 2,000 metres. If we consider the Tertiary and Quaternary 65 million years ago as representing a gigantic catastrophe and therefore only a short–term event, then not only did the dinosaurs living on the Colorado Plateau die out during a natural catastrophe only a few thousand years ago, but the Colorado Plateau itself rose out of the sea at this time, because the horizontal geological layers of the Grand Canyon were formed under water.

The many volcanic eruptions, whose lava flows according to Robert H. Webb formed lava dams which retained the water, would then fit in to this event. The water of the former lakes of the Colorado Plateau does not then come from the melt water of fictitious glaciers, because what masses of ice could cause a flood 37 times bigger than the Mississippi Flood? On the contrary, the water comes partially from inside the earth (also soil liquefaction) and mostly from the North American inland lake (see figure 8), which more or less split the western and eastern parts of America into two continents during the Cretaceous. Through the formation of the mountains (at the end of the Cretaceous = Deluge) the water was lifted up and collected in large basins. Heavy rainfall filled the lakes up again. When the dams broke, the superfloods poured out and ripped open the Grand Canyon in several phases. The water lines one can see in the dry lakes in today's deserts in the Colorado Plateau, which still seem fresh, are a reminder to the visitor of how quickly a landscape can change.

That is why today one finds dinosaur skeletons at the edge of these former lakes, like sardines in a tin, usually mixed up with crocodiles and turtles, in a type of washed–together mass grave made up of a jumble of bones. A German expedition to Tanzania (East Africa) in 1909 also found remains of Barosaurus (formerly Gigantosaurus) together with seashells, snails, belemnites and fish at Tendaguru, in deposits at a Cretaceous lake near the coast. In total, there are even three layers deposited by the sea, one on top of the other, each 20–30 cm in depth and containing dinosaur relics (*German Colonial Lexicon*, 1920, Vol. III, p. 475f.; cf. Fraas, 1909).

As discussed in detail in my German book *Dinosaurier Handbuch* (Dinosaur Compendium), most types of dinosaur did not live on land but in the water, which is why they are often found together with crocodile and turtle relics. This view previously prevailed but was scientifically superseded fairly recently. Sauropods are now considered to have been land animals which, in view of their immense weight, problems with blood supply and other criteria, cannot be correct (Zillmer, 2002, p. 87ff.).

From this viewpoint, it follows inescapably that dinosaurs could not have survived at an altitude of 2,000 metres where their remains have been found in Colorado. Definitely not! With the lifting of the complete plateau, together with the lakes and swamps in which they lived, they were lifted upwards, like in an elevator, either dead or, in isolated cases, still alive. The few surviving dinosaurs were already condemned to death and relatively immobile merely because of their size, their blood supply which was extremely difficult to maintain, and the encroaching winter impact. The surviving human inhabitants of these regions therefore saw the dinosaurs with their own eyes and drew them on the rocks exactly as they looked, and as latest scientific findings now see them: with horizontal tail, straight spine and a head which was slightly raised, at most, just like the depiction of the dinosaur on the sword excavated in Tucson in 1924, mentioned at the outset.

Anyone who visits the Navajo in Arizona can ask them about their legends. The myths bear witness to the fact that the ancestors of the Navajo and the dinosaurs lived side by side at the beginning of the world. Was that 140 million years ago or rather just a few thousand years ago? The dwellings of the Navajo close to Tuba City, Arizona are immediately adjacent to rock sheets, on the surfaces of which countless footprints of dinosaurs are preserved (topography in my German book *Dinosaurier Handbuch*, pp. 262–263). When asked, the Navajo confirm that there were also fossilised human footprints here but these were cut out of the rock sheets by the labourers who were building the road. A report published in 1990 says that not only human and dinosaur footprints are in evidence, as documented in my book *Mistake Earth Science* but that there are also fossilised traces of mammals in the same geological layers (Rosnau et al, 1990).

The footprints near Tuba City are in what is today desert, alongside fossilised heaps of dung – so-called *coprolites*: – fresh-looking but composed of hard rock, so that it appears they were left behind by dinosaurs "recently". Since that time, absolutely nothing has changed here apart from the climate, nor was anything ever buried (Photos 21–24). Can a heap of dung remain preserved on the earth's surface for 140 million years? Navajo myths also confirm that dinosaurs lived here recently. Did the dinosaurs live here in this cheerless desert? No, because this region once lay at the edge of the ancient Hopi Lake. The footprints and *coprolites* were left in the shallow shore areas of this lake, as proven by the fossilised ripple markings. Lake Hopi was drained suddenly by the Little Colorado Canyon, an erosion channel flanked by bluffs, which empties into the Grand Canyon. Soft heaps of dung and the footprints in the soft mud of the former lake remained. "Dried out" by the heat, they were then baked hydraulically into limestone by the calcium in the mud. This limestone layer is only a few centimetres thick and contains footprints, coprolites, bones and ripple markings all right beside each other in a homogeneous, undisturbed layer. This must have happened only a few thousand years ago as the *coprolites* bear no signs of erosion or weathering.

The Grand Canyon is not far from Tuba City. A signboard at the Yavapai Point Museum in Grand Canyon National Park describes the belief of the indigenous Havasupai people. According to their tradition, the earth was covered by a deluge and when the water finally receded and the mountains rose up, rivers were created; one of these cut a huge rift, which became the Grand Canyon.

These ancient Indian legends correspond exactly with the scenario described in this book: after the earth was flooded, humans saw the mountains growing and watched the waters being diverted in the shape of newly-formed rivers.

Robert H. Webb (U.S. Geological Survey in Tucson, Arizona) confirms that part of the Grand Canyon was formed by a catastrophic event during the lifetime of prehistoric Indians (cf. Fenton et al, 2002, pp. 191–215). The last phase of flooding may have been as recently as 1,300 years ago. The myths of the Havasupai appear to tell of this event. Are they also right when they say that their ancestors (and the dinosaurs) saw the Colorado Plateau rise?

But isn't everything dated precisely and the age of the stone established already? In *Darwin's Mistake* I drew attention to the basic shortcomings of radiometric dating (p. 88ff.). Using this method, one can only date eruptive rock (general types: granite and basalt), but not sediment (such as slate, sand and limestone). Fossils, however, are only to be found in sedimentary stone and therefore cannot be dated directly.

In addition, radiometric dating methods result not only in widely differing age results for granite and basalt on application of different processes, *but also when exactly the same dating method is used!* According to a scientific publication (*Arizona Bureau of Geology and Mineral Technology Bulletin*, 197/1986, p. 1ff.), the age of a lava flow in the northern area of the Colorado River was established as only 10,000 years using a potassium-argon model. A different sample from the same lava flow was said to be 117 million years old (*U.S. Atomic Energy Commission Annual Report*, NO. C00–689–76, 1967).

This very high age dating must obviously be wrong, because there was no Grand Canyon at that time. The Indian legends tell of volcanic eruptions during and after the formation of the Grand Canyon: I all happened a short time ago!

These new scientific investigations underpin, together with the legends of the indigenous Indians and the dinosaur relic finds described, that human beings lived side by side with dinosaurs not in the Mesozoic Era, but relatively recently. In this case, with a bit of luck it should be possible to find not fossilised dinosaur bones.

Fresh Traces

Not fossilised dinosaurs bones are not a rarity. In Alaska in 1961, a collection of dinosaur bones in un-fossilised and non–mineralised condition were discovered. It took more than 20 years before they were identified as the bones of duck–billed and horned dinosaurs (*Journal of Palaeontology*, Vol. 61/1, 1986, pp. 198–2009. The journal *Science* of 24 December 1993 (pp. 2020–2033) reported the finding of a sensationally well-preserved duck–billed dinosaur bone in Montana. Under the microscope it was possible to compare the well-preserved bone structure with that of a chicken.

In another case the lower jaw of a duck-billed dinosaur was found by a young Eskimo, who was working in 1987 with scientists from the *Memorial University* in Newfoundland (Canada) on Bylot Island: The bone was not fossilised and in a "fresh" condition. This find was reported in the *Edmonton Journal* of 26 October 1987 (cf. Saturday Night, August 1989, Vol. 104/8, pp. 16–19).

Further reports on not fossilised dinosaur bones are to be found in *Time magazine* of 22 September 1986 (p. 84) and in a 1982 article by Margaret Helder entitled "Fresh dinosaur bones found" which appeared in the Ex-Nihilo magazine (Vol. 14/3, 1992, pp. 16–17). Not fossilised dinosaur bones do not fit into geology's world view of the great age of dinosaurs or the corresponding age of the layers in which these bones are contained. But other old relics which should be fossilised but aren't, are also found.

Only about 1,200 km from the North Pole on Axel Heiberg Island which lies in the North Polar Sea west of Greenland, a 1995 expedition discovered a "frozen" forest, which dating showed to be 45 million years old – 2,000 miles north of the forest boundary in the Arctic. The wood is black and contains small quantities of amber, but it is not fossilised as one would expect when it has been lying for such a long time. Indeed, the wood is so fresh, that it can be cut and burned. Anyone who believes that the trees survived for 45 million years in a frozen state is working on false assumptions: There was not always ice here – in the frozen forests of the Arctic, including in Alaska – we also have the famous redwoods. These trees (sequoias) still grow today only in warmer areas like California. However, they only thrive in a damp, warm climate and certainly not in Arctic latitudes. Conclusion: the wood from these trees must have rotted over the course of millions of years, but is freshly preserved in the Arctic, as if just felled (*Time Magazine*, 22 September 1986, p. 64 and J. F. Bazinger – *Our "tropical" Arctic, Canadian Geographic*, Vol. 106/6, p. 2837, 1986/1987).

But there are other even more astonishing finds. The respected journal *Science* (Vol. 266, 18 Nov 1994, p. 1229 ff.) published an article which drew barely any attention. A dinosaur bone said to be 80 million years old was found just below the surface of a coal seam in Price, Utah. Scott R. Woodward obtained DNA from this find! For how long can DNA be preserved? Protein decays within only a few days but genetic material is supposed to be able to survive for extremely long periods of millions of years? On the basis of this discovery, Prof. Dr. Gunnar Heinsohn quite rightly asked in 1995 (p. 381) whether "they would rather insist on simple millennia (should be), where now with 80 million years impression is made".

In April 2000, scientists from the *University of Alabama* published new research findings: Tey succeeded in isolating genetic material from a Triceratops bone from North Dakota which was allegedly 65 million years old. The bone's state of preservation is interesting: it is not densely mineralised. Once one cancels or telescopes the time since the point at which the dinosaurs became extinct (Tertiary and Quaternary) to a few thousand years, these almost un-fossilised dinosaur bones are, at most, a few thousand years old. From this viewpoint, such finds become credible because DNA and un-fossilised bones only survive for short periods. This is also why human and dinosaur footprints which fossilised at the same time are not fake. Dinosaurs were still alive only a few thousand years ago, as the finds of un-fossilised bones prove.

Two separate scientific teams in America under the direction of H. R. Miller determined the age of fossilised Arcocanthosaurus bones from the Paluxy River region in Texas, using C–14 dating and a mass spectrometer (Ivanov et al, 1993). The result contradicts current evolutionary theories as the bones were found to be only 36,500 and 32,000 years old respectively. Later measurements with two different mass spectrometers produced even lower ages of 25,750 and 23,700 years (*Factum*, 2/1993, p. 46).

As dinosaurs supposedly became extinct 65 million years ago, if this dating is officially confirmed, it will amount to a paradigm shift: Goodbye to the theory of evolution on time scale grounds! To remove doubts, a project undertaken with a Russian research group made further age measurements. Using a different method, fossil dinosaur bones from north-west Siberia, the bones of modern turtles, of Cro-Magnon humans from East Kazakhstan and the Texan dinosaur bones were all dated. The co-existence of dinosaurs and humans appears to be confirmed by this project because "on the basis of the values obtained from the isotope relationships, the two dinosaur fossils could practically not be differentiated from the Cro-Magnon jaw. This means it is highly probably that both lived at the same time" (Factum, 2/1993, p. 48).

In 1997, traces of blood (!) from a Tyrannosaurus Rex at the Hell Creek Formation had already been tested, but no genetic material was found. However, such a find was reported by *Science* (Vol. 307, pp. 1952–1955) on 25 March 2005: to the surprise of palaeontologists, a Tyrannosaurus fossil from the Rocky Mountains in Montana contained numerous seemingly intact cells, well–preserved soft tissue and elastic and flexible blood vessels, after fossilised splinters were softened in weak acid. Mary Schweitzer of North Carolina State University said:

"It was totally shocking. I couldn't believe it until we'd done it (the test) 17 times". Her colleague, Lawrence Witmer from the University of Ohio agreed: "If we have tissues that are not fossilised, then we can potentially extract DNA".

Given the long periods of time postulated, until this discovery the likelihood of such finds was, correctly, thought to be zero!

Now we have to ask the question again: for how long can organic tissue remain un-fossilised?

Since Time Immemorial

Why should dinosaurs, in particular those living in water, not have survived, when a few crocodile species, turtles and sharks did? The Cologne zoologist, Ludwig *Döderlein* visited the Bay of Tokyo between 1879 and 1881. In fishermen's nets he found a rare specimen, a primeval deep-sea frilled shark (lizard shark). This is a "living fossil" with a continuous dorsal fin similar to that of an eel and which, with its eel-smooth body, was also known as a sea snake. This two metre-long animal has been alive unchanged and without any sort of evolutionary development for at least 150 million years. The teeth resemble that of the extinct genus Phoebodus which is said to have lived during the Devonian 380 million years ago. Why should this animal, which was perfectly adapted to its environment, undergo any changes? Nor was the frilled shark superseded by animals which were better adapted because this species lives in deeper regions of the outer continental shelf in the Atlantic and Pacific.

It appears that Mosasaurs also swam in the Sahara Sea until recently. The handbook *Art and Myth in Ancient Greece* (Thomas H. Carpenter, 1991), shows an urn from Turkey, dated to around 530 B. C., seems to depict a Mosasaur – together with a dolphin and *other known* sea creatures.

Sightings of sea monsters could fill entire books. In 1977, a Japanese fishing boat before New Zealand pulled up a corpse, almost 10 metres long, from a depth of 250 metres. This mysterious creature, in a state of decay, was around 2 tons in weight, had a spine, four big fins – two each at front and back – and a finless tail. The head sat on a long neck. After several photos were taken and a tissue sample extracted, which proved to be completely unrelated to shark or whale, the animal was thrown back into the water (see *Darwin's Mistake*, photos 96–99). The Japanese Post Office marked the find with a stamp depicting a Plesiosaur.

There have been reports of another monster near Victoria in British Columbia (Canada) since at least 1881. Since that time, there has been a ceaseless stream of sensational reports of a snake-like animal, up to 20 metres long; the animal has been sighted, sometimes by more than one witness at a time, 178 times. It cannot be confused with a fish or a whale, as it has no dorsal fin and its head is said to be like that of a camel. The monster has been given the nickname of "Caddy", a diminutive of Cadborosaurus. The name comes from Cadboro Bay, where the creature has often been seen.

There have been many sightings of sea monsters. But in one particular case the sightings were confirmed by a find. At a whaling station close to the Alaskan border in 1937, a freshly caught sperm whale was cut open and a largely undamaged creature, 3.2 metres long, was found in its stomach. Photographs show a thin, snake-like creature with no visible hair. There are two small front fins at the base of the neck and two further fins at the end of the tail. One eyewitness reported that the long body was covered with fur, except for the back, on which there were overlapping horny plates, covered with spikes. Nobody was able to identify the animal with the camel's head (Photo 16). The Canadian marine biologists, Paul H. LeBlond and Edward L. Bousfield (1995) gave the creature the scientific designation of *Cadborosaurus willsi*.

It seems that trees too, such as the gingko, have been in existence unchanged for 200 million years. Pollen and spores from flowers and plants have even been found in 600 million year old stone from the Precambrian (dawn of time) in Guyana (*Nature*, Vol. 210, 16 Apr 1966, pp. 292–294). At this time there was supposedly as yet no life on land, as this only appeared during the Cambrian. Is the geological time scale quite simply wrong?

These and many other finds have contributed to the fact that even some few scientists no longer accept the long periods of time, laid down for natural and human history by the theory of evolution (*Chronology and Catastrophism Review*, Vol. 15, 1995, pp. 23–28).

The existence of dinosaurs until only a few thousand years ago also makes it more understandable why some of these species are still swimming around happily in the ocean today. Thus the hypothesis is supported, that represents the Tertiary era one phantom-

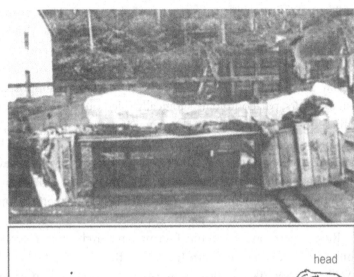

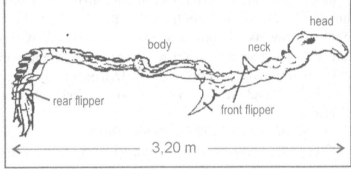

Fig. 9: Caddy. The photo shows the mysterious corpse found in the belly of a sperm whale in Naden Harbour, British Columbia, Canada: is it a Cadborosaurus? On its back there were overlapping, spiked horn platelets and its head resembled that of a camel, a description resembling those given by many eye-witnesses from other sightings of similar sea monsters in this area. The drawing (below) is a representation of the corpse. Cf. photo 15 and Baigent (1998, p. 74).

truncated age. Accordingly, the Great Ice Age, said to have lasted almost 2 million years should also be reduced to a shorter period of a few hundred years at most.

Ice Age as Time Impact

The icing over of the Antarctic is said to have begun 35 million years ago. But there are authentic old maps, which at the time they were produced in the 16th–18th Centuries or copied from older maps, show that the Antarctic was ice-free – at a time before this continent was officially discovered in 1818, for example on Philipe Buache's 1737 map. If the Antarctic already started to freeze over 35 million years ago and the inland ice is more than 30 million years old, as is assumed, our ancestors could never have mapped the landmass lying under the ice.

According to the interpretation of Jack Hough of the *University of Chicago*, which appeared in the *Journal of Geology* (1950, Vol. 58, p. 254ff.), drilling cores show "Ice Age" marine sediments covering the period from the present to 6,000 years ago. Investigations suggest that in the previous 9,000 years, i.e. 15,000 years ago, these sediments consisted of layers of fine-grained deposits sorted by size. They come from ice-free (temperate) zones and were transported by rivers to the sea. The drilling cores show that the last warm period ended at the South Pole 6,000 years ago and only then did it begin to freeze over. Is this warm period confirmed by the discovery of a fossilised fly (*Nature*, Vol. 423, 8 May 2003, pp. 135–136)? Since the Antarctic has supposedly been frozen for over 30 million years, palaeontologists logically ruled out the existence of more developed flies, to which house flies belong, in the Antarctic. There is, after all, one certainty: No heat, no flies! However, Hough had proven that there was adequate heat up to 6,000 years ago. The discovery of this fly is therefore no mystery. However, we must remove the persistent about 30 million years of Antarctic ice as a phantom time, and replace it with a period of only a few thousand years when (with the winter impact) the ice actually arrived – but very quickly (time impact).

Researchers on board the German research ship *Polarstern* discovered a region of intense volcanic activity with fresh lava flows at the Gakkel Ridge below the Arctic ice (*SpW*, 20 Nov 2001). The freezing over of the Arctic (Greenland, Spitsbergen) and the Antarctic was, inter alia a direct result of volcanism, apparently still active below the ice, and wind streams which came after the Deluge for thermal reasons. Because: No heat no rain (snowfall) and therefore no formation of glaciers. Just because it gets cold, there are no glaciers.

This phase, which I call the "Snow Time" was a sort of time-shortened "Great Ice Age" connected with the Winter Impact, (sea *Darwin's Mistake*) brought about by the impact of a large meteorite and considering that the oceans were initially warmer – explained and discussed in detail in my book *The Columbus-Mistake* (*Kolumbus kam als Letzter*, 2004). *Geology* magazine confirmed again in 2004 (Vol.32, No. 6, pp. 529–532) that a rapid and extreme cooling took place after the dinosaur impact (K/T boundary). In conclusion: the glaciers formed fast, not slowly.

The unreasonableness of postulating that the "Great Ice Age" lasted for almost 2 million years is apparent from the findings of the palaeontologist, Dr. Ralf–Dietrich Kahlke of *Jena University*, a respected Ice Age expert: He said that woolly rhino, mammoth, musk ox and bison, as well as other heat-loving animals such as elephant and lion, populated the vast regions between northern Spain and the far eastern Pacific Coast, right across the Bering Straits into North America during the "Great Ice Age". The finding is correct, but the conclusion drawn is wrong. An internet scientific report (*Informationsdienst Wissenschaft*, 26 Sept 1999) states that: "These animals bore great aridity and temperatures far below freezing point with stoic equanimity ... the permafrost zones xtended for several hundred metres deep into the earth. The character of an ecosystem is, however, defined by the duration of cold influences – millennia or thousands of millennia?" In other words: animals such as lions vegetated in the permafrost with stoic calm – without adequate food? Was this, as Kahlke suggests a perfect survival strategy or is it all just nonsense ...?

These animals can survive the onset of a cold period or a sharp drop in temperature for only a short time, if at all. Under such climatic conditions, the great herds would, in any event, find very little to eat. Having survived the long period of the "Great Ice Age" did so many animal species (an estimated 80%) really die out just at the end of the last Ice Age? This is nonsense because animals without food die when the sudden change of temperature starts and not at the end of a long period with very low temperatures. In conclusion: the ice formed fast and intensively a few thousand years ago and many animal species died because of the sudden onset of bitter cold. Latest investigations essentially confirm my opinion: 24,000 years ago, just when the Ice Age had supposedly reached its coldest point, a wide variety of flora and fauna (mammoth, bison, horse) were present in Beringia, the formerly land area of today's freezing Bering Strait (Zazula, 2003). A relatively short time ago, the "Arctic Steppe" was still a fertile grassy landscape (*Nature*, Vol. 423, 5 June 2003, p. 603).

Today, Greenland is covered in a permafrost sheet which, however, will already have disappeared within one thousand years given a rise in temperature of 3°C (*Nature*, Vol. 428, 8 Apr 2004, p. 616; cf. *Science*, Vol. 296, 31 May 2002, pp. 1687–1689 and Zillmer, 2001, p. 302). Given climatic fluctuations, ice has a relatively short lifespan, as the shrinking of glaciers worldwide shows, and it does not survive for millions of years.

A 3,028 metre-long ice-core, which was retrieved by the 1990–1992 *European Greenland Ice-Core Project* (GRIP) supposedly proves that Greenland's ice sheet is 250,000 years old. However, recognisable layering of the ice goes only 1,500 metres deep, below which there is solid ice. In the absence of layering, the age of the ice was established by measuring dust particles. It was said that at depth of around 2,300 metres, the ice was only 40,000 years old. Thus the official opinion is that the greater part of the ice was formed at the end of the Neanderthal Era, during the lifetime of modern Man (Cro–Magnon). The remaining 723 metres – around one-quarter of the ice–core – are said to represent, believe it or not, 210,000 years. As there is only solid ice at this depth, it is assumed that in the lowest part, one millimetre of ice represents one calendar year. This is unfounded speculation. How can they make this leap? Some kind of current speeds have been dreamt up and then used to construct an (unproven) ice flow model of the ice's migration speed. Then the age has been interpreted by calculating backwards. Naturally, the results obtained are directly linked to the arbitrarily-selected base criteria. If these are changed, the result will also be different.

In the meantime, it is not just these errors which have been recognised. New investigations reduce the age of this ice-core by 20% to 200,000 years at most and raise the question of whether one can draw conclusions about the period before 110,000 years ago at all, on the basis of the data obtained. The ice-core which supposedly provided a precise calendar was unexpectedly younger, simply because the scientific assumptions had changed. But even this younger age is far too old. Based on the emergency landings of eight aircraft on 15 July 1942, which were recovered from the ice 47 years later (Hayes,

1994, cf. Heinsohn, 1994), it can be proven that the ice flow model is incorrect, because at that time there was no movement at all: the aircraft. They were lying on the same coordinates they gave during the emergency landing. If we take the ice flow model and use zero as the movement value, we get an event-dependent sudden formation of ice, as outlined in my Snow Time Theory. In addition, the Ice Age experts (glaciologists) calculated that the aircraft should be covered with twelve metres of ice. In fact, it was 54 metres of ice, as well as 24 metres of extremely hard firn, i.e. 78 metres in total – 6.5 times more than predicted. Therefore, the ice grows much faster than the ice researchers believe. If one calculates the proven rate of ice coverage in the years from 1942 to 1989 to be 1.65 metres per year and transposes this rate directly onto the 3,028 metre-long ice–core (idealised marginal value calculation in keeping with Uniformitarianism), then the "perpetual" ice in Greenland is a mere 1,818 years old! Its current freezing actually started with the Little Ice Age around 1350. Before then, Greenland was green and the Vikings carried on cattle breeding and dairy farming.

The time-shortening which I already called for in my book *Mistake Earth Science* has been partially confirmed by a new ice-core drilling: The members of the *NGRIP Project* (North Greenland Ice Core Project) came to the conclusion after a fresh drilling that the lowest ice, at a depth of more than 3,000 metres, was only 123,000 years old (*NGRIP 2004 season release*, 7 August 2004). Compared with the older GRIP ice-core, this is still a reduction in age of 50%, i.e. by half. My Snow Time model suggests that there should be a further reduction in the age of the ice, if the lowest layers did not form at the same slow speed year for year but were the quick result of a natural catastrophe (cataclysm).

During the NGRIP ice-drilling which ended in 2003, the research team was surprised (Andersen, 2004), to discover a reddish coloured ice, mixed with mud above the rocky bed of the ice sheet at a depth of 3,085 metres. Embedded in this ice were brownish fir needles, tree bark and grass (*Nature*, Vol. 431, 9 Sept 2004, pp. 147–151). These plant remains are said to be several million years old. Why aren't they simply the same age as the ice sheet in which they were found, i.e. 123,000 years old? Because the trees on which the fir needles grew would have had to have been present and, according to geological time scale, these trees could only have grown on Greenland before the start of the "Great Ice Age" – because of the (supposed) ice blanket over Greenland.

This is a circular argument because according to the ice model, the freezing of the Arctic began (20 million years later than the South Pole), 10–15 million years ago, but the ice-cores were, at most, 0.2 million years old. That scientists constantly come up with such hair-raising misinterpretations is due to the fact that they repeatedly fall victim to their own statistical models and the resulting circular arguments. If we shorten the supposed geological timeframes, these interesting investigations become more plausible: Was Greenland ice-free at this time, as has been proven for the Bering Straits (*Nature*, Vol. 423, 5 June 2003, p. 603) at the time of the allegedly coldest point of the last Ice Age 24,000 years ago? This point in time of ice-free Arctic regions corresponds in my model to the ice–free phase before the Deluge up to 5,500 (or perhaps 4,500) years ago.

It is important to note that the research team correctly sees the plant remains under the ice and above the bedrock as a sign that the ice formed rapidly. I have to underline this sensational finding! The rapid, and not slow, freezing, especially of mountains, is exactly one of my fundamental theses. But one can only come to this conclusion if the plant material is still fresh and not primeval. Or do leave and grass remains fresh for hundreds of thousands of years, until the sudden onset of freezing? Surely not, because leaves rot quickly! So let's forget these arbitrarily inserted millions of years, let the landscape freeze over quickly and thus allow fresh plant remains to be preserved: officially. This would be 123,000 years ago according to the dating of the more recent NGRIP ice-core and not 10 million years ago by geological dating, or by my Snow Time model, it happened at most only 5,000 years ago after the Deluge – Greenland, after all, means "Green Land".

The Venetian brothers, Nicolò and Antonio Zeno, who themselves travelled the North Atlantic in 1380, produced a map but it was only published for the first time in 1558 in Venice by Francesco Marcolino. The expert, Prof. Charles H. Hapgood (1996) of *Keene State College* in New Hampshire, notes that this map shows Greenland free of ice with mountains and rivers, in polar projection. He also found that the topography below the ice in Greenland corresponds with that shown on the map. In other words, Greenland was mapped in ice–free conditions. This is why I had predicted that modern human relics, only a few thousand years old, would be found beneath the "perpetual" ice. With regard to Greenland, my statements have already been proven because, to the astonishment of the scientists, a Viking farm was found, more or less under the ice, at Nipaatsoq: Soil investigations have shown that life would have been impossible on Greenland's more northerly coasts in the middle of the 14th Century, because of the Little Ice Age, according to Charles Schwager, Professor of Archaeology at the *University of Alberta* (*New York Times*, 8 May 2001). Excavations directed by Jette Arneborg have since unearthed two thousand artefacts, all of which indicate that the Vikings packed up calmly and left their settlement. Archaeological analyses, soil samples and pollen tests show that it was not – as was believed for a long time – warfare which led to the abandonment of the settlement but, in fact, a change in climate (*BdW*, 10 May 2001). In other words, the Vikings saw the ice approaching the farm, which was then buried by the glacial sand.

In any event, these centuries-old maps (Portolan maps) prove that science's traditional world view is most certainly wrong, as modern Man appear to have seen "perpetually frozen" regions free of ice, which they then measured and mapped. Conclusion: The "Great Ice Age" is a fiction because the "perpetual ice" arrived rapidly during the course of the Snow Time, as a secondary effect of the Deluge, melted with the greenhouse climate during the Roman Climatic Optimum and the warm period in the first century A.D., and then formed again in the Arctic (Greenland and Canada) during the Little Ice

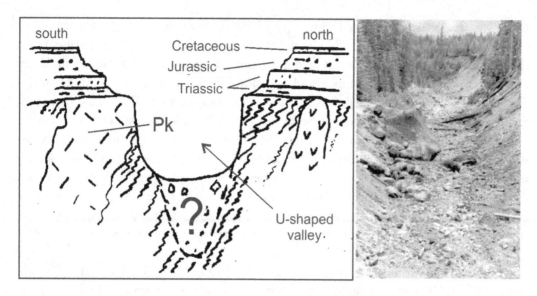

Fig. 10: Misinterpretation. *V-shaped valleys are formed through very long-lasting erosion by smaller masses of water (creek, river). Science therefore considers that U-shaped valleys must have been formed by the scraping action of a glacier. However, large masses of water also (or perhaps only) create U-shaped valleys. The cross-section sketch shows the U-shaped Unaweep Canyon in Colorado, which was formed by large masses of water. Untold tons of erratic blocks were left behind. The photo shows a smaller U-shaped erosion channel, formed by mud flows, which were thrown out during the Mount St. Helens eruption in 1980. Pc = Precambrian (the dawn of earth).*

Age around 1350 A.D. If, as previously discussed, the period after the dinosaur impact (Tertiary and Pleistocene) only existed as a time-shorted period, then there is something radically wrong with the evolutionary history of mammals and humans, given that these supposedly developed during a period after the dinosaurs became extinct. But there are also other proofs which contradict human evolution.

2   A Jumble of Bones

Henry Gee, the publisher of the well-known scientific journal "Nature", refers to evolutionist intrigues in relation to the evolution of human beings as a completely human invention created subseuquent, shaped to accord with human prejudices, and he adds, to take a line of fossils and claim that they represent a lineage is not a scientific hypothesis that can be tested, but an assertion that carries the same validity as a bedtime story – entertaining, perhaps even instructive, but not scientific (Gee, 1999, p. 126f.) The evolution of Man from ape-like ancestor is a modern, fantastical fairy story – with no basis in truth!

Trees, Apes and Hominids

The skeletons of humans and apes differ radically from each other. Apes live predominantly in trees and are therefore quadruped. Humans, on the other hand, are characterised by their upright posture. Evolutionists insist that this means of locomotion evolved out of movement on four legs. In 1996, the anatomist, Prof. Robin Crompton proved with a three-dimensional computer simulation, that bipedality cannot evolve out of quadrupedity. The result contradicts accepted teachings. The way in which Lucy, an (allegedly) ape–like ur-ancestor of Man, moved, upright, with bowed back and bent knees, was shown to be non-viable. Scientists are therefore convinced that our ancestors must always have walked upright (cf. Sarre, 1994 and Deloison, 2004) or that they became extinct or changed from four legs to two legs, within the shortest possible space of time, even before they left the trees, contradicted Crompton at the annual meeting of British scientists 1996 (*Journal of Human Evolution*, Vol. 31, 1996, p. 517–535; cf. Henke, 1996).

Conclusion: A living creature moves either upright *or* on four legs. The above investigation shows that a method of locomotion which lies in between these two movement types (half bi-pedal) is impossible and, on logical grounds alone, is improbable, because the missing links required to support this argument are neither to be seen in the animal kingdom nor have they been found in fossil form.

In contrast to the theory of evolution, it is more likely that former bipeds change over to moving on four legs, if they are forced to do so by external circumstances – like to life in trees, in swamps or in steep mountains (cf. Sarre, 1994 and Deloison, 2004).

The statement I made in my 1998 published German book, *Darwin's Mistake*, to the effect that humans have always gone upright and that his ancestors cannot have lived in trees for anatomical, statistical and logical reasons, have now found support. The anthropologist, Carsten Niebuhr from the *Freie Universität Berlin*, tends to assume that our

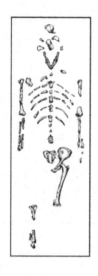

Fig. 11: Lucy. These bones were the basis for the reconstruction of our "first mother", Lucy, representing countless generations of creatures who walked upright.

ancestors lived largely on the ground: wading upright, our early ancestors looked for food in shallow waters and during the day, they walked the savanna; only at night did they climb into trees, where they could sleep in safety. Because of the diversity of habitat, there was no specialisation. Man did not become a specialist at walking, even though he has a characteristic foot for walking, which is suited to moving over large distances using little energy, for running fast over short distances or for swimming. Similarly, *Homo sapiens* did not develop his prehensility to a high degree. Niebuhr contradicts "the theory, that we differ from animal primates in that we have a 'creative hand' for fine manipulation" by pointing out that the hands of apes are anatomically adapted to their specific needs ... the hands of apes (are) much more 'modern' than those of the human. The human hand is comparatively primitive in comparison; it is mainly our brain which makes the human hand superior" (*BdW*, 31 Jan 2001).

Current research therefore shows that there has been no specialisation in the shape of advanced development with regard to the functionality of Man's extremities. Bipedality, in particular, is not an evolutionary advantage, because the locomotive mode of the ape is far simpler, faster and more effective than the upright posture of humans. Since the human does not have any great prehensility, one might actually speak of a retrograde step in development. The superiority of humans over ideally-adapted animals has nothing to do with the evolutionary developments of the skeleton, but in the application of intelligence, controlling the body of an "all rounder" which allows the human to appear to be a superior specialist.

Lucy, The Ape

What about our alleged ancestor, Lucy who, representing the ape species *Australopithecus afarensis* was said to have lived 3.6 million years ago? The assertion that Lucy went upright is, in fact, a view which was successfully and media-effectively represented for decades by paleoanthropologists such as Richard Leakey and Donald C. Johanson (Johanson/Edey, 1981).

The ape-headed Australopithecines allegedly have some adaptations for bipedality, particularly in the area of the pelvis and the lower extremities. This is why the Australopithecines (Australomorphs) are seen as the missing link between (Miocene) great apes and Man. However, in Lucy's body the centre of gravity is not between the hips, but higher and further forward, making walking even more difficult than it already is.

Detailed research studies by several scientists on the skeletal structure of Australopithecines have been carried out. Lord Solly Zuckerman and Prof. Charles Oxnard, two world-renowned anatomists from the USA and England, undertook extensive research on various Australopithecus specimens which showed that these creatures could not walk with Man's upright posture. After studying the fossil bones for 15 years, Zuckerman and his team came to the conclusion that Australopithecus was a species of ape and could certainly not be bipedal (Zuckerman, 1970, p. 75ff.).

In accordance with this opinion Charles E. Oxnard integrate the bone structure of Australopithecus in the same category as that of modern orang-utans (*Nature*, Vol. 258, p. 389). According to kinematic tests, Lucy could not walk normally, i.e. was not statically stable (Crompton et al *in Journal of Human Evolution*, 1998, Vol. 35, pp. 55–74).

Conclusion: Australopithecines are not connected with humans. They are merely an extinct species of ape and do not represent a link in human evolution.

Having analysed the semi-circular shaped channels of the inner ear in apes and humans, whose function is to maintain a sense of balance, anatomy experts came to a similar conclusion: The dimensions of the channels in Australopithecus and his successor (Paranthropus) are very similar to those of today's ape (*Nature*, Vol. 369, 23 June 1994, p. 645ff.).

The evolutionists' argument too, that chimpanzee DNA is almost identical to that of humans because of a common ancestor, is a propaganda lie. Because not only the statics of the skeleton but also the differences in DNA between humans and chimpanzees are much greater than previously thought. The chimpanzee and the human genome differ not by 1% as previously claimed, but by five percent (*PNAS*, 15 Oct 2002, Vol. 9, pp. 13633–13635 and 15 Apr 2003, Vol. 100, pp. 4661–4665).

Flexible Anatomy

The skull attachment in humans and apes is fundamentally different and demonstrates, at the very least, separate paths of development. During an evolutionary phase of development from quadrupedal to bipedal locomotion, the change in body posture would also have required comprehensive changes to the entire skull. Evolution theorists are also aware of these anatomical differences, but officially ignore them. Naturally, because if one has no reasonable explanation, discussion of the issue would do more harm than good and the inadequacy of evolution theory would be manifest.

In moving to an upright posture, the centre of gravity of the skull attachment would have to gradually change as well, from the rear edge of the ape's skull to the middle of the human skull. The foramen magnum, where the spine connects with the skull, should therefore have migrated, so to speak, along the skull, because the chimpanzee's skull literally hangs on the spine, whilst in humans, the skull sits on the spine. This significant anatomical difference cannot have been caused by pure coincidence or a freak of nature. There are no fossil records of any of the missing links which would have been required for the ape's successful move to an upright position. To date, only one or the other skull attachment type has been found, but nothing in between.

Where the human can point and finely grip with his hand, the chimpanzee tends to make a fist. He cannot hold objects between his thumb and forefinger. The gelada, which belongs to the baboon family, is actually more talented with its hands and fingers than the chimpanzee! Apes have the ability to grip equally well with all four hands and feet, which is why scientists accurately coined the term "quadrumana" (four–handed). Looking at the morphology of the human foot bones and their anatomy, it is clear that the human foot is solely designed for locomotion and not for hanging in trees. How could the human foot bones have changed so completely that they can no longer grip? Is this an evolutionary step forwards – or better perhaps, backwards – in terms of evolution theory? Why didn't the prehensile foot change in such a way as to allow both gripping and upright locomotion? Or are we talking about two completely separate anatomical features with no connection or development?

Shouldn't the position of the leg bones in modern humans and chimpanzees also be similar if both have a common ancestor, as is categorically asserted? Using visual analogies and similarities, one is tempted to answer this question with "yes", but the facts tend the other way. It is at least clear, that the ancestors of humans did not come down from the trees, they are not descended from apes, nor do they have a common ancestor – unless the ancestor was himself biped. In other words, would be the only possible development that humans and apes are descended from an ancestor that walked on two legs. Only later evolved apes, and have nothing to do with the Human History.

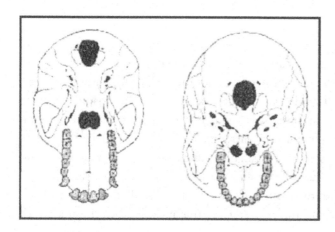

Fig. 12: Foramen magnum.
The ape's head hangs on the spine (left) whilst in humans its centre of gravity rests on the spine (right).
From the schoolbook "Biologie heute S II", 1998, p.425.

The Missed Rendezvous

Mutated chromosomes are not usually suitable for reproduction, as they are rejected by the intact ovum (egg cell). A human cannot breed with a chimpanzee and equally, under natural conditions, animals of different species cannot reproduce with each other. What does it actually mean, to belong to the same genus or the same species? Using this nomenclature, which goes back to the Swede, Carl Linnaeus (1707–1778), a living organism is allocated two Latin names: the first is the name of the genus, which means a group of related species, whilst the second is the designation of the species, which is defined as the totality of those individuals capable of producing offspring which may in turn bear fertile offspring.

Burros and horses, for example, both belong to the same genus (Equus), but not the same species, because the offspring produced from their interbreeding – hinnies and mules – are infertile.

I have photographed rare offspring on the Galapagos Archipelago, which was produced by the crossing of the marine iguana and the Galapagos land iguana. Strangely, the interbreeding of these two species only produces offspring if the father is a marine iguana and the mother a land iguana. The offspring (hybrids) are, in any event, always infertile; they can therefore not propagate and thus also cannot climb higher on the imaginary ladder of evolution. New research shows that land iguanas genetically separated from marine iguanas over 20 million years ago (BdW, 15 July 1999) and therefore allegedly had common ancestors. Human offspring would be similarly infertile, if there had been progeny during the transition from an ape species to a Homo species.

Genetically mutated animals and humans seem to become isolated because if we suppose for a moment that an imaginary hybrid were capable of reproduction, it would still require a sexual partner: a male needs a female (and vice versa) with a correspondingly altered set of chromosomes in order to reproduce. That means not only must a single animal mutate, but the partner required for reproduction must also (co–)mutate in the same way at the same time.

Geographically speaking, how did Adam and Eve ever manage to find each other? For it is of little use if Adam lived, for example in northern East Africa and his correspondingly mutated Eve in southern Africa or perhaps even on a different continent. There is also a problem of time, which I have already mentioned. For an evolution, it is senseless if Eve is alive with mutated chromosomes but Adam is already dead, or vice versa.

Latest genetic investigations carried out by Peter Underhill's team show that the "Adam and Eve" of modern Man missed each other by around 80,000 years because according to this genetic analysis, Adam didn't exist until 59,000 years ago (*Nature Genetics*, Vol. 26/3, November 2000, pp. 253–254 and 358–361).

Today, two different tools are used to calculate genetic family trees. Since a spermatozoon (sperm cell) does not pass on its mitochondria (allegedly) when it melds with an ovule, most of the mitochondrial DNA (mtDNA) comes from the mother's egg cell. Using mtDNA analyses the maternal ancestry can be calculated back to the "original Eve". The "original Adam", on the other hand, is calculated using the Y-chromosome. Since this is transmitted only from fathers to sons, one can establish male ancestral lines. These family trees should then more or less correspond. But, as Peter Underhill's investigation shows, this is not the case.

Geneticists such as Luigi Luca Cavalli-Sforza reject the idea of an "original Eve" because if one uses this data, there is no proof at all that there was any time in history when the human population was reduced to a single woman or that at the time of this so-called Eve the population density was particularly low (Cavalli-Sforza, 1999, p. 101). One must agree with him, as long as one is speaking of one and the same species, e.g. humans reproducing amongst themselves (microevolution). However, there must have been an original Eve and an original Adam if, through a kind of transcendent mutation, a human should evolved from an ape-like creature (macroevolution). Like all orthodox evolution theorists, Cavalli–Sforza does not make a distinction here. In order to deceive, the facts of microevolution are used as evidence of macroevolution, because the change above the level of species is unproven and is therefore speculation.

Let's take a closer look at the genetic calculation on the basis of mutation rates. On the basis of constant frequency of mutation, one finds that chimpanzees and humans had a common ancestor around five million years ago. Does this prove common ancestry? Surely it can't be proven with a mathematical formula. The genetic distance between two populations, regardless of whether big or small, which transcends the species boundary, does not prove that there were common ancestors. Furthermore a constant frequency of mutation per time unit must be seen as an abstract conceptual model only.

Today's mutation rates are known: they are very slow, relatively linear (i.e. predictable) and regionally limited. Thus the regions and periods of time of change are calculable. The result is time-spans of over 130,000 years for the evolution of proven mitochondrial DNA (mtDNA) of modern Man, which represents the age of the "scientific Eve". A new investigation has however cast doubt on the mutation rate used for this calculation. Comparing DNA sequences over a period of 327 generations, an empirical rate was established, which is shorter by a factor of 20 than was taken as the basis for the previous phylogenetic analyses (*Nature Genetics*, Vol. 15, 1 Apr 1997, pp. 363–368).

According to this investigation, the "scientific original Eve" would be only 6,500 years old, rather than 130,000. If one considers the rejuvenation factor, the first settlement of America may have taken place only 1,500 years ago rather than 30,000 and the last pre–historic settlement of America only 500 years ago, rather than 10,000. And finally, the separation of chimpanzees and humans would have occurred only 250,000 years ago rather than 500,000. Time-spans thus become significantly shorter and more tangible.

Another point which favours the argument that humans have existed for only a short time, is that all humans are genetically more similar than was previously thought (*Science*, Vol. 294, 23 Nov 2001, pp. 1719–1723). If humanity were old, there should be greater genetic differences. Is this not the case because humanity was once almost completely wiped out (*PNAS*, 1999, Vol. 96, pp. 5077–5082)?

For all humans living today to be directly related to Adam and Eve, one would need only 33 generations with an average reproductive age of 25, i.e. a total of 825 years. If, with this statistical calculation, one also takes factors such as geography, history and migration into account, our identical ancestor lived less than 5,000 years ago, i.e. 169 generations (Rohde, 2004). This means, that one person 5,000 years ago was either the ancestor of all people living today or his line became genetically extinct. He cannot have been the ancestor of only a few people living today. Using a computer, a research team created several scenarios, integrating factors such as varying population growth, isolation of individual groups, local migration and mass migration. The results of this study show that our most recent common ancestor probably could have lived 3,000 years ago (*Nature*, Vol. 431, 30 Sept 2004, pp. 562–566). Has modern Man (including our Stone Age ancestors) been in existence for only a few thousand years?

The population of India is set to explode from one to two billion people within only 20 years and, in 45 years, it is predicted that the world's population will grow by 50%, i.e. from 6 to 9 billion. This makes it clear how quickly developments can occur.

Fictional Family Trees

During the Mesozoic Era 65 million years ago, at the time of the dinosaurs, there were allegedly only small primitive mammals which were only able to develop further after the dinosaurs became extinct. The first primates are thought to have appeared at the end of the Eocene (55–36 million years). According to Evolutionism, the tree-dwelling New World apes (e.g. howler monkeys), Old World apes (e.g. proboscis and rhesus) and the great apes (gibbons, orang–utans, gorillas and chimpanzees) developed one after the other. The transitions between these groups are, like all transitions from one animal species to another, not confirmed by any fossil finds.

For a long time, Ramapithecus who was dated to between 14–8 million years before Present (B.P.) was seen as one of the early stages on the road to Mankind. In the 1960s he was even considered to be Man's ancestor. The reconstruction of Ramapithecus was based solely on two pieces of an upper jaw. In the book *Early Man* (Howell, 1969) it was firmly asserted that: "if one superimposes the teeth and palate of Ramapithecus firstly onto the palate of an orang–utan and then onto the palate of a human, the human resemblance is obvious..." Today, it tends to be seen the other way around. The once supposedly human features are now said to most closely resemble those of the orang–utan. It must be emphasised that Evolutionists classify Ramapithecus as a separate transitional form, based solely on a few teeth and part of a palate. Many generations containing millions of individuals and entire sections of family trees were wrongly invented. Is this a regrettable error or a deliberate manipulation?

"On the basis of a few teeth and part of a palate, it was claimed that Ramapithecus resembled a human. The proof was the supposedly curved palate, which curved outwards as in humans. Apes and great apes on the other hand have U–shaped flat palates" (Howell, 1969).

I don't wish to go into the early segment of the (alleged) human family tree in more detail because there is no evidence to confirm man's early ancestry; therefore going into detail would be imaginary and purely speculative. Furthermore fossil ancestors have not been documented for either chimpanzees or gorillas. This is why the corresponding family trees comprise only straight trees with no connections, as not a single missing link has been found. Our knowledge is fragmentary. The older part of the family tree up to Australopithecines (from the Latin for "southern ape") is imaginary, as there is a vast lack of corroborative finds.

The next step of human evolution, classified as Homo (*Latin*: man) allegedly came after the Australopithecines. According to Evolutionist claims the creatures of the Homo-series are more highly developed than the Australopithecines. Until recently, *Homo habilis* (2.4–1.5 million years B.P.) was regarded as the link between Australopithecus and the true human being. The first specimen was discovered with tools and was somewhat prematurely classified by science as *Homo habilis* (skilful Man). After this, the first human is said to have appeared in the form of *Homo erectus*. Scientists have widely varying opinions on *Homo erectus* (*Latin*: upright Man). Some differentiate between the earlier Homo ergaster and the later *Homo erectus*. Others differentiate between the African Homo ergaster, the Asian *Homo erectus* and the European Homo heidelbergensis, previously known as the archaic *Homo sapiens*, who is now said to have been preceded by Homo antecessor. In between there are various other designations to choose from which, in one form or another, have at one time or another been regarded as the standard designation – usually intensely disputed by the experts.

Such so-called hybrids (e.g. Steinheim Man) and early-modern Cro-Magnon Man supposedly led to the development of today's modern humans. Based on the DNA studies carried out by the geneticist, Prof. Dr. Svante Pääbo, some scientists have now ruled out Neanderthal Man as an ancestor of modern humans (*Science*, Vol. 277, 1977, pp. 1021–1025). Now the Neanderthal is regarded as either an extinct sub-species of modern Man (*Homo sapiens* neanderthalensis) or (since his discovery and currently in the USA) as a distinct species (Homo neanderthalensis).

A distinct species would mean that Neanderthals and modern Man could not have reproduced together.

In any event, the Neanderthal is a discarded link in the chain of human evolution, for which no replacement link has been found. But, say the Evolutionists, some other link

Fig. 13: Dwarfism. *The siblings of the Owitch family were born of dwarf parents. If a paleoanthropologist in the distant future were to discover the bones of a tall basketball player, he could assume that humans in the 21st Century were extremely tall. If, however, he were to discover the bones of the Owitch family, it would appear that we were small bipeds.*

will be found to take the place of the Neanderthal. Of course, those who believe in Evolution miracles, can simply pull another rabbit out of the hat, in the shape of a missing (imaginary) Homo–ancestor, who is waiting to be discovered somewhere in the ground in the darkness of time. Perhaps another jaw fragment will be found and this will be declared to be the proof of a species which takes the place of the Neanderthal and will by proxy represent millions of undiscovered individuals. By the time, years later, that it is again established that this was the jaw of an ape, the idea that it was a missing link will have taken firm hold in people's minds.

In declaring the evolutionary chain to be Australopithecus – *Homo habilis* – *Homo erectus* – Home sapiens – it means that each of these species was the predecessor of the following species and, according to Darwin's laws, all levels must have existed one after the other. However, latest discoveries by some paleoanthropologists reveal that Australopithecus, *Homo habilis* and *Homo erectus* lived at the same time in different parts of the world. In addition, some proportion of the humans classified as *Homo erectus* lived on into the modern age. *Science* reported in an article headed "Latest *Homo erectus* of Java: Potential Contemporaneity with *Homo sapiens* in Southeast Asia", that *Homo erectus* fossils had been dated to an average age of 53,300–27,000 years. So *Homo erectus* lived together with anatomically modern Man (*Homo sapiens*) in South East Asia (*Science*, 13 Dec 1996, Vol. 274, pp. 1870–1874)!

It is said that for around 3,000 years, only modern man (*Homo sapiens*) has survived as the sole representative of the human genus, since the extinction of the Neanderthal. Is this assumption wrong as well? In September 2003, the complete skeleton of a previously unknown human species was discovered on the Indonesian island of Flores and was given the new species name of Homo floresiensis, said to have lived only 18,000 years ago (*Nature*, 28 Oct 2004, Vol. 431, pp. 1055–1061 and 1087–1091).

Was this really a modern human being rather than a great ape (compare photos 48–49). After all, relatively advanced implements were found beside the skeleton. But the remarkable Flores Man was supposedly not a specimen of *Homo sapiens* because it was characterised by its extreme dwarfishness of only one metre in height and the brain volume, amounting to 380 cubic centimetres, was tiny. Even Lucy, (*Australopithecus afarensis*) had a greater brain volume. It was only the anthropologist, Maciej Henneberg of Adelaide University who complained: "The dimension knocks everything I have learned about evolution in 32 years on the head" (*Focus*, 10/2005, p. 153). Until then, it had been assumed that hominids of such small physical stature and brain, such as the newly–discovered Flores human, had already died out as an evolutionary link 3 million years ago.

The consistency of the un-fossilised bones of Flores Man are described as "like mashed potato". Can such bones remain preserved for 18,000 years in an un-fossilised condition? Together with the bones, remains of the still extant Komodo dragon and a dwarf elephant of the Stegodon family, said to have become extinct on Flores 5,000 years ago, were found. The dig workers speculate that Flores Man fell victim to violent volcanic eruptions. Why not? He sought refuge in the cave and was then buried beneath a total of six metres of mud (tsunami?). Under these circumstances, Flores Man might be only 5,000 years old – or perhaps significantly younger?

Scientists see Flores Man (Homo floresiensis), as a direct ancestor of *Homo erectus*, that Java Man discovered by Eugene Dubois in 1891. According to the human biologist, Prof. Günter Bräuer (*University of Hamburg*), if the Flores species really did develop out of *Homo erectus*, their separation must lie even further back than the currently propagated separation of modern Man and Neanderthal, 600,000 years ago (*Die Welt*, 31 Oct 2004, p. 76). Where then are the countless individuals belonging to this development chain?

The alleged Flores "Hobbit" recalls the Proto–Pygmies, a classification created by the cryptozoologist, Ivan T. Sanderson which contains all smaller hominids (wild people) such as the Orang Pendek, Sehite, Teh–Ima and Agogwe pygmies. These unknown hominids exist more or less all over the world, if eyewitness reports are to be believed. The pygmies are familiar with the Agogwe, but according to their myths, these shy jungle dwellers were the first creatures born at the dawn of time. This is why the pygmies honour the Agogwe as forest spirits and leave them gifts, for which they in turn sometimes give something back: by leading lost children back to camp or helping in some other way.

The well-known cryptozoologist, Bernard Heuvelmans suspects that the Agogwe are a graceful species of Australopithecus which survived in East Africa. Around 1900, the lion hunter, Captain William Hitchen observed two small men, of around 1.2 metres height, with brown fur and upright posture, in the area of Wembere, Tanzania as he lay in a hide during an official lion hunt (cf. Bord, 1995, p. 416). This report was not published until 1937 by the *Discovery magazine* whereupon Cuthbert Burgoyne responded saying that from a boat, he had seen similar creatures in 1927 in East Africa. It was believed that these were Agogwe.

The footprints of the Agogwe are around 12.5 centimetres in length and the big toe is, in comparison with the others, bigger than in humans and spreads sideways. Agogwe have long fur, either rust-coloured or, as has occasionally been reported, ranging from black to grey. The skin is more yellowish to light brown and they have relatively long arms. The height varies between 90 and 120 centimetres and corresponds to the size of Lucy. Are there still living specimens of Australopithecus or, if the reports are correct, is this an as yet unknown species of human?

One repeatedly hears reports from Tanzania and northern Mozambique of so-called Agogure or Agogue, a human–like creature resembling the pygmies but with longer arms. They have longer reddish fur. There are reports of similar creatures in Guinea, Sierra Leone, Côte d'Ivoire and in the Congo Basin. At Ngoliba in Kenya, it is said that other "hominids, probably survivors from prehistoric times, have been demonstrably observed in eleven forest areas by 33 indigenous people" (Bord, 1995, p. 397). On the other side of the Atlantic too, in Colombia, hominids of 1.2–1.5 metres in height, known as Shiru have been sighted on various occasions.

The fossils which are purported by the Evolutionists to be Man's forerunners are, in fact, either different hominid species or various species of ape. It appears that these creatures lived until only a short time ago, or that they even still exist in inaccessible regions of the earth. However, this coexistence of species is a diametrical contradiction of Darwin's principle of selection.

Conclusion: the ancestral series of Man posited by the Evolutionists represents an arbitrary lining up of a few bones and skulls. The material basis for human evolutionary history for perhaps 250,000 generations in around 4 million years from the start of development with *Australopithecus afarensis*up to the Neanderthal is based on only about 300 bone fragments which can be ascribed to fewer than 50 people. In other words, there is one, largely fragmentary find for every 3,000 generations. Here, the phylogenesis of modern Man falls down in several instances, because according to simulations, the first ancestor of modern Man can not have lived longer than 169 generations ago (*Nature*, Vol. 431, 30 Sept 2004, pp. 562–566).

How can one produce entire lines of development and ancestral hypotheses at all, in the face of these discrepancies and with such isolated finds, and then also present them as scientifically proven facts? In this case, belief is surely more important than science. Almost every new find tangles the alleged ancestral chain again. At the moment, there is a trend towards moving Man's beginnings backwards in time to the age of the dinosaurs, which is starting to coincide with the idea of coexisting humans and dinosaurs which I have postulated. But on the other hand, this trend further stretches the already very loose chain of human fossil finds.

The Relegation of Links

The human evolutionary line is also disputed by scientists and does not represent a proven fact. Since the transition from chimpanzee–like Australopithecus to *Homo erectus* with his much greater brain volume seems too erratic and does not document a slow evolutionary process of development, a transitional form was required. *Homo habilis* therefore had to fill the gap in order to lend credibility to evolution theory.

Some scientists however, including Willfred Le Gros Clark, question the existence of the transitional link *Homo habilis*. Loring Brace wrote (Fix, 1984, p. 143): "*Homo habilis* is an empty taxon inadequately proposed and should be formally sunk". J.T. Robinson even insisted that these funds were a mistaken mixture of skeleton parts from Australopithecus (ape) and *Homo erectus* (Man). Louis Leakey asked whether the term *Homo habilis* did not, in fact, include two species of the Homo genus, one of which continued to develop into *Homo sapiens* while the other became *Homo erectus* (Wood, 1987, p. 187).

A skull and skeleton fossil, known as OH 62, discovered by Tim White and postulated to be *Homo habilis*, was used as proof for the existence of a transitional form. However, the reconstruction revealed a small skull capacity, long arms and short legs: overall the characteristic features of Australopithecus, an ape. In 1994, the American anthropologist, Holly Smith published a detailed analysis and pointed out that *Homo habilis* was not a human (Homo), but an ape (American Journal of Physical Anthropology, Vol. 94, 1994, p. 307ff.).

Furthermore, a specimen of *Homo habilis* (STW 53) was examined and was found to be even more ape–like than the Australopithecines and was even more poorly equipped to walk upright. The conclusion: "STW 53 probably does not represent an intermediate stage between Australopithecines and *Homo erectus*" (*Nature*, Vol. 369, 23 June 1994, p. 645ff.).

Because of the very heterogeneous morphology, following the find of skull KNM-ER 1470 in 1972 (Leakey, 1973), *Homo habilis* was finally split into two species: *Homo habilis* and *Homo rudolfensis*. Richard Leakey, who excavated the fossil, presented the allegedly 2.8 million year old skull KNM-ER 1470 to the public as the greatest discovery in the history of anthropology. According to Leakey, this creature (*Homo rudolfensis*), which like Australopithecus had a small brain (cranial capacity), but the facial structure of a human, was the missing link between Australopithecus (ape) and Man. Only a short time later, however, it was established that the human facial structure of skull KNM-ER 1470, which had often appeared on the front pages of scientific journals, was the result of the incorrect assembly of the skull fragments: in reality, the face looked even more like an ape than Australopithecus (*New Scientist*, Vol. 133, 11 Jan 1992, pp. 38–41).

The paleoanthropologist, J.E. Cronin stated that the relatively primitive characteristics of the alleged missing link *Homo rudolfensis* connect him with the members of the taxonomic ape genus Australopithecus africanus (*Nature*, Vol. 292, 1981, p. 113ff.). Conclusion: *Homo rudolfensis* is not a human, but an ape. Loring Brace from the University of Michigan came to the same conclusion after an analysis he carried out on the jaw and tooth structure of the skull ER 1470. He said that this find, celebrated as a missing link, had exactly the same face and teeth of an Australopithecus (ape) (Brace et al, 1979). The paleoanthropologist Prof. Alan Walker also insists that the celebrated skull KNM-ER 1470, as a species of Australopithecus, must be regarded as an ape (Scientific American, Vol. 239/2, 19778, p. 54).

When KNM-ER 1470 was found, the skull was said to be 2.9 million years old, as it had been found beneath a tufa layer which had been radiometrically dated to an age of 2.6 million years. As was later established, the age of the tufa had been wrongly calculated, because the samples examined had been contaminated with older volcanic stone. The age of KNM-ER 1470 had to be reduced to 1.8 million years – making it probably exactly the same age as *Homo habilis*! Therefore an allegedly old link in the evolutionary chain was seemingly transformed into a younger link in another (imaginary) chain.

In brief, we can state that classifications such as *Homo habilis* or *Homo rudolfensis*, which are represented as missing links between the Australopithecines and *Homo erectus*, are entirely imaginary – i.e. these missing links do not exist in fossil form. As has now been confirmed by numerous scientists (Science, Vol. 284, 2 Apr 1999, pp. 65–71), these creatures are a type of Australopithecus – in other words, apes.

The missing links between humans and apes which, according to evolution theory must be widely present in fossil form, do not exist. This new finding is only starting to gain prominence now, following decades during which *Homo habilis* was held up and celebrated in scientific journals and magazines as proof of human evolution and the correctness of evolution theory. The previously postulated missing links have proven to be a "Fata Morgana". In the meantime, *Homo habilis* has been reclassified by some scientists as Australopithecus habilis (*Science*, Vol. 284, 1999, pp. 65–71); in other words, he has been pushed back down the evolutionary ladder. On the basis of new morphological evidence, *Homo habilis* introduced by Louis Leakey in the early 1960s has lost his status as a human being. The link position between ape and human is orphaned. Does it even exist? Comparison of the skulls of *Australopithecus afarensis* and a modern-day ape reveal almost no differences.

It is therefore not surprising that Charles E. Oxnard of the University of Chicago writes (1975, p. 393 f.): It is rather unlikely that any of the Australopithecines can have any direct phylogenetic link with the genus Homo. As a genus (Australopithecus fossils), there is a mosaic of unique features and characteristics, which resemble those of an orang-utan.

It appears that Lucy was a type of orang-utan. As apes always remain apes and there are no missing links, let's now take a look at the human species.

Homo Species

The evolution of the human being (Homo species) should have occurred as follows: *Homo erectus*, archaic *Homo sapiens*, Neanderthal (disputed), Cro–Magnon Man (early-modern man) and finally the modern human. In fact, what we are talking about are simply various species of modern man. The differences between them are no greater than those between pygmies, Eskimos and Europeans.

The cranial capacity of *Homo erectus* fluctuated between 900 and 1,100 cubic centimetres and is therefore at the lower margin of modern Man's cranial capacity. Let us firstly examine *Homo erectus*, who is represented as a primitive form of human being. The Evolutionists allowed *Homo erectus* to differ from the great apes by giving him the quality of "upright posture". But there is no difference between the skeleton of modern Man and *Homo erectus*.

The main reason why the Evolutionists classified *Homo erectus* as primitive, was firstly his brain size (900–1,100 cubic centimetre), which is smaller than that of the average modern human being and secondly his heavy overhanging brows. There are, however, people living today who have the same size brain (cranial capacity) as *Homo erectus* (e.g. pygmies) and there are many other peoples, such as the Australian Aborigines, who have prominent brows.

The fossils with which *Homo erectus* became known to the world were *Peking Man* and Java Man, who were found in Asia. Over the course of time however, it became clear that both of these fossils were doubtful, as I shall discuss later. As a result, *Homo erectus* fossils found in Africa were accorded greater significance. Some alleged *Homo erectus* fossils were categorised by some Evolutionists in another class, called Homo ergaster. All of these fossils are discussed below under the category of *Homo erectus*. The most famous example of *Homo erectus*, which was found in Africa, is the Turkana Boy, found close to Lake Turkana in Kenya. The upright skeleton posture does not differ from that of a modern human being. Speaking about this, the American paleoanthropologist, Alan Walker, said he doubted whether an average pathologist would be in a position to establish any differences between the fossil skeleton and a modern human – and the skull "almost exactly resembled that of a Neanderthal" (*The Washington Post,* 19 Nov 1984).

Fig. 14: Comparison. *Not extinct. The skull of* Homo erectus *compared with a living man (right). In both cases, the prominent browline and other features such as the receding forehead, are recognisable, and these features exist in a number of peoples today. Photos: Harun Yahya.*

Even the Evolutionist, Richard Leakey, stated that there were few differences between *Homo erectus* and modern humans. Prof. William Laughlin of the University of Connecticut carried out detailed anatomical examinations on Eskimos and inhabitants of the Aleutian Islands chain. He noticed an extraordinary similarity between these people and *Homo erectus*. Conclusion: in all of these instances, including *Homo erectus*, we are dealing simply with different variations of modern humans (Lubenow, 1992, p. 136).

The view that *Homo erectus* is a superfluous classification is gaining ground in the scientific world. The *American Scientist* reported that delegates to the *Senckenberg Conference* in 2000 became embroiled in a heated debate about the taxonomic status of *Homo erectus*. The debate was sparked off by the contention presented by several scientists, that *Homo erectus* could not be considered as a distinct species and should be dropped altogether: *Homo erectus* is *Homo sapiens*. All representatives of the Homo species, from around 2 million years ago through to the present day, are a largely mutable, widely distributed species, *Homo sapiens*, without natural interruptions or sub-species (*American Scientist*, November/December 2000, p. 491). This, the view of some scientists, exactly corresponds with the opinions presented in this book.

As such, the genus name erectus becomes redundant, simply because erectus and sapiens are one and the same species. Perhaps François de Sarre is right when he said that *Homo erectus* represents a different, "wild" form of human. To use an analogy, the skeletons of brown bears and polar bears barely differ but they have two completely different lifestyles. If one considers that *Homo habilis* and *Homo erectus* no longer exist, there is now an unbridgeable chasm between apes (Australopithecus) and man (*Homo erectus*, *Homo sapiens*): in terms of fossil finds, the first human appeared suddenly with no evolutionary pre-history.

This fact completely contradicts the dogmatic philosophy and ideology of the Evolutionists. Reconstructions of *Homo erectus* are therefore created with ape-like facial features whilst, on the other hand, apes such as Australopithecus and *Homo habilis* are artistically humanised. Thus the chasm gaping between these different, clearly separated hominid classes is artificially bridged. The illusion becomes reality in the public consciousness. The best-preserved skeleton of *Homo erectus* is probably the find designated KNM-WT 15000. This fossil is said to be 1.6 million years old and actually presents evidence against evolution theory in that it resembles the far younger Neanderthal. The Evolutionist, Donald Johnson, even compared the shape and proportions of the specimen's limbs to today's equatorial Africans.

Prof. Helmut Ziegert of *Hamburg University* discovered 400,000 year old traces of a settlement and 200,000 year old remains of roundhouses at the edge of a prehistoric lake the size of Germany, in the Sahara (*BdW*, Issue 4/1998, p. 18ff.). Long before the Neanderthals were around, these early humans (allegedly *Homo erectus*) were already producing specialised tools. "The early human used boats and fished, hunted ostrich and wore leather clothing". Ziegert therefore says: "I oppose the reconstructions which show early humans half–naked or wearing skins". Conclusion: *Homo erectus* lived like a modern human.

It is actually the Neanderthal who should be considered the successor of *Homo erectus* and the forerunner of modern Man. But genetic investigations seem to suggest that the Neanderthal was not our ancestor. This is why the "Case of the Neanderthal" is discussed in a separate chapter. In any event, there is no replacement for the Neanderthal as the forerunner of modern humans in Europe: the modern human suddenly appeared in Europe out of the depths of history around 45,000–35,000 years ago.

To make up for the Neanderthal, the point in time at which modern humans first appeared has been pushed further back in history. When two modern skulls (Omo 1 and 2) were found by Richard Leakey in Ethiopia in 1967, the age of the skulls was established as 130,000 years, based on the radiometric measurement of the uranium and thorium decay in volcanic layers. Since the Neanderthal had recently been ruled out as our ancestor, it was felt that new measurements of the previously dated volcanic ash had to be undertaken. This time, the radioactive decay of potassium and argon was used and, lo and behold, the intended back–dating of modern humans was published as a scientifically proven fact: Man is now about 200,000 years old (*Nature*, Vol. 433, 17 Feb 2005, pp. 733–736). Mission accomplished because now the gap left behind by the Neanderthal is bridged. However, by dating the first appearance of humans in Africa to so far back in time, the rift between the first anatomically modern human and the first clear signs of cultural development has widened, because for at least 150,000 years of this period not even the tiniest find has been uncovered.

Human Co–Existence

The disputed time impact during the Pleistocene (Diluvium) related to the geological layers has an anthropological parallel, since the first Homo species entered the stage of human history because allegedly primitive human species and more highly developed great apes lived together with modern humans, although according to Darwin, the first group mentioned were replaced by more highly evolved types.

However, "There is evidence from East Africa for late-surviving small Australopithecus individuals that were contemporaneous first with *Homo habilis* and then with *Homo erectus* (*Science*, Vol. 207, 07 Mar 1980, p. 1103). Louis Leakey found fossils of Australopithecus, *Homo habilis* and *Home erectus* almost right beside each other in the Olduvai Gorge region in the Bed II Layer (Kelso, 1970, p. 221).

Stephen Jay Gould, a well known palaeontologist at Harvard University explains the evolutionary dead end as follows: What has become of our ladder if there are three coexisting lineages of hominids (*A. africanus*, the robust australopithecines and *Homo habilis*), none clearly derived from another? Moreover, none of the three display any evolutionary trends during their tenure on earth (*Natural History*, Vol. 85, 1976, p. 30). The co-existence of these missing links between ape and human contradicts evolution theory which states that they must have existed successively. Man cannot be descended from an ape.

When we move further to the human species, from *Homo erectus* to modern Man (*Homo sapiens*), equally there is no stepladder: all of the alleged "missing links" lived at the same time. Indeed proof exists that *Homo erectus* and a forerunner of modern humans – the so-called archaic *Homo sapiens* were still alive 27,000 years ago and possibly even 13,000 years ago. In the Kow Swamp in Australia, 22 individuals were discovered with ages ranging between 13,000 and only 6,500 years. The discoverer, Alan Thorne, gave his opinion in *Nature* (Vol. 238, pp. 316–319) that these finds represented an almost completely preserved form of the eastern *Homo erectus*, which clearly differs from today's Aborigines (Thorne/Macumber, 1972, p. 319). In other words, the supposed first stage of human development coexisted with Man a few thousand years ago (ibid., p. 316). These allegedly "primitive" humans were bigger than today's Aborigines, had extremely large heads with robust features whilst their bodies were characterised by heavy bones and powerful muscles. However, exact findings were never published. The Evolutionists did not wish to accept the fact that only 10,000 years ago or less, modern Man was still coexisting with a "primitive" species (*Homo erectus*), who looked exactly like their *Homo erectus* forerunner a million years ago.

On the other hand, fossils of modern humans are being discovered in geological layers which are far too old, i.e. where there should supposedly only be primitive human forerunners (*Homo erectus*). Paleoanthropological data revealed that representatives of *Homo sapiens*, who looked exactly like us, lived up to a million years ago, in other words about 850,000 years too early. The famous paleoanthropologist, Louis Leakey, was the first to make such discoveries. In 1932, Leakey found some fossils from the middle Pleistocene, in the Kanjera region which surrounds Lake Victoria in Kenya. This epoch, however, occurred a million years ago (*Science* News, Vol. 115, 1979, p. 196f.). Since these discoveries turn the evolutionary family tree upside down, they were not officially recognised by the Evolutionist paleoanthropologists.

This controversy was revived in 1995 when a human fossil was discovered in the karst caves of the Sierra de Atapuerca in northern Spain. With an age of 800,000 years, however, the find was 200,000 years too old, given that it had previously been assumed that *Homo erectus* first migrated to Europe 600,000 years ago. But the real sensation is that while something "primitive" had been expected, what was actually found "was a completely modern face" (*Discover Magazine*, December 1997, p. 97ff.). This skull possessed a pit-like indentation, at the lower edge of the cheekbone, the so–called fossa crania. This pit is an important morphological feature of modern humans. No other type of human, neither *Homo erectus* nor Neanderthals possessed this fossa cranium.

The shock for the anthropologists was that the history of *Homo sapiens* would have had to have been pushed back 800,000 years. As this would have contradicted human evolutionary history, a new fantasy species was invented and given the name of *Homo antecessor* (Man forerunner) and this is how the Atapuerca skull was classified.

Can modern Man be as much as 1.7 million years old? At the beginning of the 1970s, Louis Leakey discovered a stone hut, the construction style of which may still be seen today in some parts of Africa, together with fossils of Australopithecus, *Homo habilis* and *Homo erectus* in the so–called Bed II Layer in the Olduvai Gorge. Does this find indicate that modern man has been in existence for 1.7 million years? Or, based on the collocation of these finds, should one not rather conclude that instead of the hundreds of thousands of years which have been estimated, the real figure should not exceed millennia or centuries? Given these extraordinarily sparse finds, one can really only bring a few generations back to life.

"More recently, fossil species have been assigned to Homo on the basis of absolute brain size, inferences about language ability and hand function, and retrodictions about their ability to fashion stone tools. With only a few exceptions, the definition and use of the genus within human evolution, and the demarcation of Homo, have been treated as if they are unproblematic... in practice fossil hominin species are assigned to Homo on the basis of one or more out of four criteria... It is now evident, however, that none of these criteria is satisfactory. The Cerebral Rubicon is problematic because absolute cranial capacity is of questionable biological significance. Likewise, there is compelling evidence that language function cannot be reliably inferred from the gross appearance of the brain, and that the language–related parts of the brain are not as well localized as earlier studies had implied..." (*Science*, Vol. 284, 2 Apr 1999, pp. 65–71).

There will be further finds, new species and classifications which will cause a snakes and ladders effect on the hominid ladder and repeatedly throw the imaginary family tree into confusion. In these cases, however, we are speaking of either an ape or a modern human, as there are no extant fossils to document a transition from ape to man. When finds are made, which contradict the propagated family tree, they are not taken into account, because there were apes who could go upright long before the supposed first upright ape, Lucy.

The upright mountain ape

Hardly known but therefore all the more interesting is the fact that around 6 million years before Lucy, there was an ape which went upright. As long ago as 1872, François Louis Paul Gervais (1816–1879) described the "Mountain Ape of Bamboli" (*Oreopithecus bambolii*). Since then, there has been controversy as to how this primate can be classified in the order of animals (Engesser, 1998, p. 2; Brandt, 1999, p. 33), even though there is no scarcity of find material for this ape. Numerous fossil bones, including a complete Oreopithecus skeleton have been discovered in Upper Miocene lignite layers, which are around 10 million years old (Herder Lexicon, 1994, p. 250). All of the finds were made in Tuscany and Sardinia (Italy).

Current opinion holds that the last common ancestor of man and chimpanzee lived around 5–6 million years ago. Going by conventional dating, however, Oreopithecus lived 9–7 million years ago. But this must mean that Oreopithecus's human-like characteristics and his upright posture ceased to exist in later forms and only appeared again during the process of human evolution.

What is sensational about this find, therefore, is that it has only now been recognised that long before Lucy, who toddled rather than walked, there were already biped apes. This is proven by the anatomical features of spine, thighs and feet. Nor is Oreopithecus seen as a forerunner of humans. Naturally, because there are no further missing links in evidence to cover a period of several million years. This absence of evidence means that this find is no longer officially discussed. Furthermore, it is exactly man's ability to walk upright which supposedly differentiates him from the ape – a longstanding and obviously falsely propagated dogma.

Conclusion: It is compellingly shown by the case of Oreopithecus that bipedality does not necessarily indicate a forerunner of humans (Brandt, 1999, p. 36). This is why the aquatic ape theory is also discussed; this theory suggests that Oreopithecus often had to cross water in his habitat. His upright posture was surely helpful, as it allowed him to wade securely through deeper water. To what extent this aquatic environment was responsible for the evolution of Oreopithecus's bipedality, is itself an interesting question. It should not be overlooked, however, that Oreopithecus's upper limbs were longer than his legs, tending to make them unsuited for swimming. Therefore any adaptation to water can only have been limited. In addition, the long arms and curved phalanges (finger bones) are an indication that Oreopithecus would certainly have climbed trees on a regular basis (Feustel, 1990, p. 53; cf. Engesser, 1998, p. 4). Perhaps an indication of flooded rainforests?

In any case it appears clear that Oreopithecus's upright posture existed independently of the upright posture of the Australopithecenes and the Homo genus – if the geological age datings are correct. Moreover, going on evolution theory, Oreopithecus must have forerunners. As is the case with all the rest, these evolutionary stages do not exist. The question is whether parallel evolutions characterised by countless coincidences and mutational leaps can occur repeatedly i.e. that walking upright could have developed at different times in different creatures independent of each other.

Aquatic Apes

Oreopithecus's bipedality can be connected with his watery environment. Before we explore this idea more closely, let's take a look at whether Man's forerunner could have developed as a "savanna animal human", now that we have seen that an evolutionary phase of development in the trees is an untenable theory.

According to orthodox savanna theory, the size of the forests shrunk dramatically around 4 million years ago because of climatic changes. Because of the diminishing supply of food, the primate population was under pressure, so a few daring individuals went to find food outside these forests in the savanna. They moved out into the great, grass-covered plains of Africa. Natural selection supposedly then led to "modern" human characteristics proving themselves to be advantageous. Whilst these characteristics evolved slowly, the human rose up on his hind legs, so that he could see over the high grass. At the same time, his brain grew and his heavy pelt disappeared.

At the outset, it was already discussed that an evolution from crawling to walking upright cannot have occurred (inter alia in *Journal of Human Evolution*, 1998, Vol. 35, pp. 55–74). Alongside his upright walk, the human also has other physical features which differentiate him from apes and other land mammals. Unlike these, the human can breathe equally well through nose or mouth. On the other hand, humans are unique in that they cannot drink and breathe simultaneously. The reason for this is a particular physical feature: the low position of our larynx.

In land mammals the mouth is connected with the stomach via the oesophagus, whilst the nose is connected to the lungs by the windpipe. Unlike humans, these animals can therefore breathe and drink at the same time. The windpipe runs through the palate by means of a ring-shaped sphincter. When this sphincter relaxes, air can be expelled from the upper end of the windpipe into the oral cavity and equally be sucked in by the windpipe. This process is responsible, for example, for the barking of dogs. After barking, the windpipe lifts again and the sphincter closes. The separation of windpipe and oesophagus is restored.

In humans, on the other hand, the windpipe is under the base of the tongue, i.e. lower than in other animals. Humans also do not possess a sphincter to separate windpipe and oesophagus. This is the only reason why both food and air can reach the lungs, as well as the stomach, because the rear side of our palate is open. This is why, in humans, swallowing can be dangerous, because food or liquid can enter the windpipe. This is also why choking whilst eating is a fairly common cause of death amongst humans as compared with other mammals.

Biologists are completely baffled as to why this unusual biological construction should have developed through natural selection during the transition from forest habitat to savanna (Morgan, 1990, p. 126). All experts agree that this biological feature is truly unique. There is also no definitive explanation for its origins. But in this case too, there can not have been any evolutionary stages from one physical structure to the other. To use an analogy: an advanced machine only works when its construction has been thoroughly thought out and it has been perfectly built. Errors in planning or construction lead to standstill or breakdown, in exactly the same way that an incompletely developed oesophagus or windpipe would lead to the death of the creature in question. Slow evolutionary steps in man or animal would of necessity lead to the death of the species. Such a development is not possible and this is why fossil and contemporary evidence show only consistently examples of "Intelligent Design" – "thought out" biomechanisms perfect from the start and fully adapted to their respective environments.

Just as little can a lower-positioned larynx have been of evolutionary advantage in the savanna. Apes and other land mammals manage perfectly well with the other system. So we have to ask the questions: under what conditions would the human's lower–positioned larynx be of advantage? Could it be if humans lived in a watery environment such as that in which the upright mountain ape lived?

The human also has a remarkably thick layer of fat beneath his skin. Whilst this is absent in land-living primates, over 30% of human fat is distributed subcutaneously. This layer of fat is the norm for aquatic mammals. It is an excellent insulation against the loss of body heat – although when close to the equator only in water (Morgan, 1990, p. 47). Is this the reason why humans are not hairy? It is still entirely unexplained why humans lost their body hair. In the savanna, being naked is surely not an evolutionary advantage but rather a glaring disadvantage.

Walking upright, the unique method of breathing and the absence of body hair are, however, decisive advantages if living close to the sea or flooded terrain (Morgan, 1990, p. 47). Walking upright has hardly any immediately apparent advantages on land, but in water, land–living enemies can be evaded and breathing above water, either swimming or wading, is possible.

"The location where Lucy was discovered already suggests a possible affinity with water. It seems that it was once the swampy, perhaps wooded shore of a lake. Lucy's bones lay amongst the remains of crab claws and crocodile and turtle eggs" (Baigent, 1998, p. 129). Lucy supposedly drowned because she went to the swamp to drink. Primitive cultures in Africa never live at the water's edge, because these watering holes are visited by dangerous predators. Could Lucy have lived part of the time in the water? Seven million years ago, Lucy's northern Ethiopian home was allegedly a large inland lake, which gradually dried out over millions of years, leaving behind a salt plain of several hundred metres in length. Did this evaporation scenario perhaps occur at the same time as the Sahara was formed, only 5,000 years ago? Did certain ape and human species live in the flat regions of the once broad seas, before intense volcanism and the consequent desertification took place?

Dramatic Changes

Where do these contentions bring us? I would now like to offer you a brief overview of the possible events which took place during the Deluge (end of the age of dinosaurs), to give the interested reader an insight into the themes and arguments of my other books, so that a full overview is guaranteed. To avoid repetition, I have chosen to forgo any further detailed argumentation at this point.

As already demonstrated by O.C. Hilgenberg (1933), Klaus Vogel (1990) from Werdau has been showing with his increasingly impressive globes, some of which are made of glass, that all of today's continents, including the now submerged continental shelves, fit together astonishingly well to form an "original earth" which was much smaller than it is today. The earth's surface now comprises 70% sea and 30% land. If one imagines the continents – like the 5 and 6 panels of a leather ball – joined into a closed globe (super continent, Pangaea with no oceans!), then we have an "original earth" the surface of which is 100% land mass, but which then has a radius which is 35–40% smaller. This small earth has only one crust, which completely encircles the earth, and no oceans in between.

Off the cuff question: Where's the water? Possible answer: above the supercontinent. The water covered and encircled the earth completely and formed a kind of greenhouse steam atmosphere above it (cf. *Darwin's Mistake*, p. 152ff.). After all, over 95% of fossils come from the sea.

The physicist and astronomer, Heinz Haber (1965) gave this formation the Greek name of "Panthalassic" ("all seas") Earth, which has hills, but no high mountains.

If we imagine a smaller earth, we see that the ocean basins were originally not as deep as they are today. If, for the purpose of this exercise we also imagine that the total volume of seawater has remained constant, then of necessity, when the earth's expansion began, a much greater part of the continental shelves must have been part of the flat–lying underwater shelf region than is the case today. In other words, what are now dry depressions must once have been under water. After examining two different geological charts, L. Egyed developed the two curves shown at Fig. 15 (*Geologische Rundschau*, Vol. 46, 1957, p. 108 and Vol. 50, 1960, p. 251). They show that the earth must once have been a planet entirely covered with water, especially when one also considers the smaller earth diameter. At the time of the dinosaurs, the earth was far more extensively covered with water (cf. Fig. 8, p. 31).

In this connection, the earth expansion theory is underpinned by oceanographic echo

Fig. 15: Under water. Water coverage of present land mass in the geological past according to Egyed. In earlier times a large part of today's land mass was under water. From Jordan, 1966.

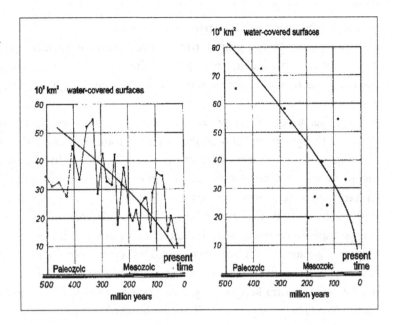

soundings, which show there are only small masses of sediment at the bottom of the sea. At present, erosion is so strong that, with the same strength in the past, an entirely different situation should have arisen. In other words: the oceans should contain vast masses of sediment. Correspondingly, with regard to erosion, instead of a growth of the continental shelf, a simple expansion of the land surface is possible, in that surfaces once covered by water became dry land.

"Finally it should be mentioned, that those sediments which are more than 2,000 million years old, contain hardly any sandstone. This appears understandable when one considers that only high lying, small continents were above the surface of the water at that time, so that neither large desert areas nor wide beach areas were available for the formation of sand by means of the sorting of stone rubble" (Jordan, 1966, p. 60).

Only after the previously discussed abrupt tilting of the earth's axis by around 20° to its present elliptical inclination at the end of the age of dinosaurs (Deluge), did the mountain chains (Himalayas, Andes, Alps) rise up to their present heights. The seemingly fresh, unweathered seashells found in these mountains at altitudes of several thousand metres prove that, as Charles Darwin already established, the mountain chains still largely formed part of the sea bottom only a relatively short time ago.

After the tilting of the earth axis (True Polar Wander), the greater part of the earth's expansion also took place (discussed in detail in *Mistake Earth Science*, p. 71ff.). At this time, the deep-sea rifts were also formed. Through the tension in the earth's crust caused by the expansion of the globe and the tilting of the earth's axis, tears were inevitably formed in the weaker sections of the earth's crust (incl. the East African Rift Valley, the Rhine Rift). The same took place in the northeast German Basin: supposedly at the end of the Carboniferous "the space was stretched; the earth's crust became thinner and enormous quantities of granite rock poured onto the surface" (Bayer, 2002, p. 31). Tears were, however, particularly prevalent in the thin ocean floor: the mid–oceanic ridges and the deep–sea rifts were formed. It is officially believed that the 11 km deep Marianas Rift was formed only six million to, at most, 9 million years ago (*Science*, Vol. 305, 4 Feb 2005, p. 689), i.e. in earth's more recent past. If one takes these allegedly long geological periods as a chronological unit, this scenario took place almost during Lucy's lifetime.

As all of the continents were closer to each other before these events, one can explain the presence of dinosaur species on the "wrong" continents, even during the Upper Cretaceous, which represents an unsolvable puzzle to geophysicists who normally have nothing to do with dinosaur paleogeography. At the end of the Cretaceous there were, for example, tyrannosaurs and their close relatives who existed only at the end of the Upper Cretaceous. They populated several continents and islands (including Madagascar), which is said to have already been separated by deep oceans a long time before – in contradiction to plate tectonic theory (discussed in detail in my book *Dinosaurier Handbuch*, p. 54ff.). Today, the earth is only expanding by centimetres, as is shown by the fact that the continents are still moving further apart from each other. Before the rising of the mountains there were considerably more flat water zones on the earth. Were early humans, Lucy's relatives and the upright mountain ape all inhabitants of this other, watery environment with large–scale flat water zones?

Our mountains are supposedly up to 3,900 million years old. Every day, erosion material in the shape of rubble, gravel, sand and clay is transported by the rivers to lakes and oceans. If we calculate a constant rate of sedimentation during earth's history, the hills and mountains cannot be more than 15 million years old because, after this period, all of the earth's elevations would have been eroded by the destructive effects of the elements (earthquake, wind, frost, water). This period must be reduced if it can be proven that erosion was more intense. Similar considerations, measurements and calculations can be made with regard the mud deposits on the floor of Alp lakes. It was established that these lakes could not be more than a few thousand years old, as they would otherwise have been filled in long ago by the deposits of sand, gravel and clay. Similar calculations show that Niagara Falls are 7,000 years old, if one uses the erosion rate of 1.5 metres per year, a figure established in 1764 (Wolfe, 1949, p. 176). Taking an increased drainage of water following a natural catastrophe into consideration, an even younger age of 2,500 to, at most, 4,000 years, is obtained (Velikovsky, 1956/1980, p. 177).

As such, one can also calculate the median rate of increase in the oceans' salt content in the past. The rivers would have needed 62 million years to carry the present salt content to the oceans. Was there no saltwater at the time of the dinosaurs? In addition, the deltas of every single river in the world are too small. Only 5,000 years would have been necessary for the formation of today's Mississippi Delta, if one assumes a constant

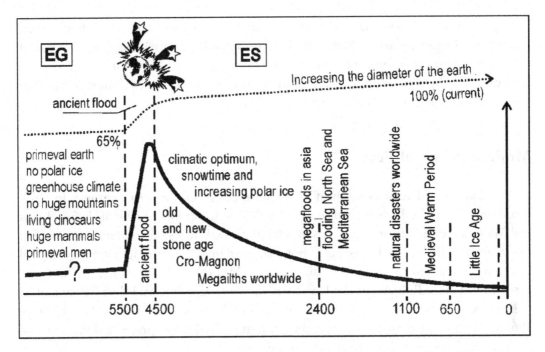

Fig. 16: Earth Expansion. *The expansion of the earth's diameter has led, according to Wegener, to the continents moving apart from each other. Before this event (the ancient flood) the earth's axis was relatively straight (EG) and afterwards slanted (ES).*

sedimentation rate. A short lifespan is also indicated by the older river deltas some of which are today below sea level.

Northeast of the lake Chiemsee in Bavaria, there is an ellipse–shaped scatter field with 81 impact craters, caused by fragments of a disintegrating meteorite. In these craters, which are up to 370 metres deep, lakes such as the Tütten Lake have formed, which were created only around 2,200 years ago during the meteorite spectacle at the time of the Celts. Seemingly very old land formations are often very young.

At the time of our forefathers, the existing lakes – such as the German Lake Constance or Ammersee, Federsee and Swiss lakes rose sharply – with a concomitant formation of barrier beaches and shoreline terraces and an accompanying destruction of all pole constructions and other shore settlements. At this time of climate deterioration earth crust movements reached a particular intensity and led to the formation of new lakes in Munich, Tölz and Memmingen (Germany). Sand drift and loess formation reached an end during this period and the dunes at Lake Constance, the Upper Rhine and in other regions subsequently became forested. A scientific study has shown that these earth crust movements in the Alp regions took place at the time of the Celts in the Sub–Atlantic period (official) from 850–120 B.C. (Gams/Nordhagen, 1923, p. 304f.).

69

Popular myths all over the world say that the mountains rose, the great deserts were formed, the Grand Canyon was dug out by erosion and the Rift Valley was torn out of the earth's crust, while intense volcanic eruptions scattered layers of ash across the landscape. At the time – geologically speaking before Lucy, modern humans left their footprints in the region of Man's supposed evolution.

Modern Primeval Feet

The find site of Laetoli is south of the Olduvai Gorge in northern Tanzania. In 1979, members of Mary Leakey's team discovered fossilised footprints of animals before Louis and Mary Leakey and their youngest son, Philip, noticed traces of hominids preserved in the layers of volcanic ash. By means of potassium–argon dating, the age of this geological layer was established to be 3.8–3.6 million years (M. Leakey, 1979, p. 452). A footprint expert, Dr. Louise Robbins of the University of North Carolina said that: "They looked so human, so modern, to be found in tuffs so old" (*Science* News, Vol. 115, 1979, p. 196f.). Mary Leakey herself wrote that this creature walked fully upright and the shape of its "primeval" foot exactly corresponded to our own.

This find, which captured the public's imagination, and the accompanying conclusions, throw up a serious problem. We now have to ask ourselves who this ancestor is supposed to have been? According to the human ancestral line, the only species which went upright 3.6 million years ago was the ape Australopithecus, if we assume that evolution chronology is correct. However, if an ape-like forerunner of Man left the Laetoli footprints, they should look different, because unlike in modern humans, the big toe of apes alive at that time was clearly separated. This anatomical difference is proof of a different lifestyle, because arboreal animals have to be able to grip with both hands and feet, whilst the human foot is in no way suited to tree dwelling. Therefore, no ape–like forerunner of Man can have left those footprints. But why couldn't it have been a modern human, if he supposedly possessed "human and modern–looking feet" (*Science* News, Vol. 115, 1979, p. 196f.). Because this would indisputably prove that modern humans did not appear only 140,000 years ago, but lived at the same time as their supposed ancestors, 360,000 years ago. To put it clearly: under these circumstances, there can be no history of humanity with a hominid family tree. This would however also confirm that modern humans already existed millions of years ago, without leaving any other traces.

All of these possibilities appear questionable. So what solutions are there to this problem? If modern humans are not millions of years old and are found in a certain geological layer, then that layer must be correspondingly young, since it contains human remains or footprints. This once again brings my proposed solution of time telescoping as discussed in my books, into play.

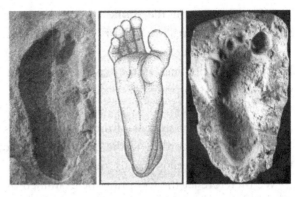

Fig. 17: Footprints. *The footprints from Laetoli in East Africa (left), which are allegedly 3.6 million years old, must correspond to the level of development of the ape (Australopithecus) at that time. However, they look as modern as the footprint of a Neanderthal in an Italian cave (right). For comparison, a 44.3 cm footprint of an allegedly 15 million year old giant ape in weight-bearing posture was reconstructed (drawing): the typically spread big toe of ape and giant ape would have to be present in the footprint of Australopithecus as well. Picture centre: Prof. Holger Preuschoft, Picture right: Howell, 1969.*

The layer of ash–drift created by the volcanic eruptions can, therefore, not be older than 3.6 million years, but can only have been deposited during an eruption just a few thousand years ago. This would solve the problem: the footprints were left relatively recently in a still–soft layer of volcanic ash after an eruption. Volcanic ash hardens quickly and not over long geological periods. On this basis, this geological layer represents a period which must be measured in days rather than millions of years. Almost all finds of ape skulls, supposedly millions of years old, were found in similar volcanic layers which were wrongly dated as older than they really are. The question of dating volcanic layers is explored in more detail in the next chapter.

If my postulated theory on the coexistence of modern humans with their alleged ancestors is correct, then one should find the various evolutionary stages of the hominid in the same geological layers. Dr. Dubois, the discoverer of *Pithecanthropus erectus* (*Homo erectus*), also unearthed in the same layer in which he found the earliest human, two human skulls, which resemble those of today's Australian Aborigines. It is instructive to note that these finds were kept under lock and key for 20 years and no official position was taken on them. Had the existence of modern skulls in the same geological layer as *Homo erectus* fossils been scientifically documented, there would most certainly have been bitter consequences for a whole series of theories.

If modern and supposedly primitive skulls lie together in the same geological layers but artefacts and fossilised human bones are found in layers which are several hundred million years old, there cannot have been a (macro)evolutionary development. As this biological timescale is directly and inextricably linked with the geological timescale, one could telescope the time horizon, i.e. shorten the period of time, without having to replace the geological period. Under these conditions, supposedly primitive and modern humans must have coexisted. This is why one today still finds seemingly primitive peoples who live in the Stone Age and whose skulls have receding foreheads. Is a receding forehead really a sign of primitiveness?

The case of the Laetoli footprints clearly demonstrates that linking geological and biological chronologies as the basis for evolution theory doesn't work. Absolutely not! Just as human footprints are declared to be fake if they are found to date back to the time of the dinosaurs, so must the Laetoli footprints also be fake, because their modern appearance can be only tens of thousands of years old at most, and not millions of years. However, since according to the experts the geological layer has been indisputably dated to the time of Australopithecus, this is clear evidence of the falseness of evolution theory.

We therefore let the footprints be examined by critical experts, so that no accusations of deliberate manipulation can be made. Louise Robbins from the *University of North Carolina* said after examining the footprints: The arch is raised … and the big toe is large and aligned with the second toe … The toes grip the ground like human toes. You do not see this in other animal forms (*Science News*, Vol. 115, 1979, p. 196f.). Russell H. Tuttle, who also examined the footprints, wrote: "A small barefoot *Homo sapiens* could have made them … In all discernible morphological features, the feet of the individuals that made the trails are indistinguishable from those of modern humans" (*New Scientist*, Vol. 98, 1983, p. 373).

The footprints were also examined by Don Johanson and Tim White. The latter wrote: They are like modern footprints. If one were left in the sand of a California beach today, and a four-year old were asked what it was, he would instantly say that somebody had walked there. He wouldn't be able to tell it from a hundred other prints on the beach, nor would you (Johanson/Edy, 1981, p. 250).

According to these exports, the allegedly 3.6 million year old Laetoli footprints can not be distinguished from those of modern humans. Where then is the evolution from ape foot to human foot? Wasn't there any? The Laetoli footprints are supposedly 200,000 years older than Lucy's but are anatomically identical not with hers but with those of modern humans. Accordingly, modern humans must have lived before Lucy and Australopithecus is not a forerunner of humans. The late Prof. Lord Zuckerman, a prominent zoologist, had already presented this viewpoint at a lecture before the Zoological Society of London. With regard to the general state of the debate, he remarked critically that contrary to evidence, the voice of authority had spoken and its (false) message was now firmly anchored in all the textbooks.

The modern appearance of the Laeotoli footprints was therefore not made public and discussed. A mountain of research funding aside, the reputation of many scientists depends on the importance of Lucy for Mankind's evolution being preserved. This is why Lucy's supposed role is still seen as a recognised fact and many experts who know better are still keeping quiet and clinging on to this manufactured dogma.

Russell H. Tuttle (1990, p. 61ff.) is right: In any case we should shelve the loose assumption that the Laetoli footprints were made by Lucy's kind, *Australopithecus afarensis* (cf. *Science*, Vol. 166, 1969, p. 958).

Strangely, all experts and laymen accept the dating of the volcanic ash in which the fossilised Laetoli footprints were found. This raises the question of when the volcanic eruption really took place, because this would also be the time of modern humans. After an alleged 3,800,000 years these fossilised Laetoli footprints are still only a few centimetres below the current surface of the earth. This example exemplifies in full the dilemma of making the link between geology and evolution sacrosanct. However, if the volcanic ash containing the Laetoli footprints is younger, or even significantly younger than dated, it shows a very important reality. If we assume that the Laetoli footprints are only a few thousand years old, there is no problem with the presence of modern humans at the time of the eruption. The riddle would be solved. However, this solution would mean that Lucy really only lived a few thousand years ago as well, during the age of modern humans.

Rare Bones

Footprints such as those described at Laetoli only remain preserved if, firstly, the layer in which they were made was soft and secondly, if this soft layer then hardens rapidly. Nowadays if one leaves a footprint in mud, it disappears within a short space of time, eroded by environmental influences (rain, wind). This means that geological layers containing footprints must have hardened fast. Ash–drift from a volcanic eruption fulfils this condition, unlike sand or mud. Preserved footprints in ash–drift therefore represent "a one-day event" from the footprint being made in the layer to the layer's hardening, and including the crossing of the area by individuals. No long geological periods of time or ancient layers are necessary.

Bones only remain preserved for a period of more than a few thousand years if they have been kept airtight or have been specially preserved. One must remember that today, not a single bone of a single animal in nature fossilises, and most certainly not near the Equator, although numerous animals die there every day. This is a circumstance which earth and human history researchers, geologists and geophysicists like to overlook because if they were to acknowledge this fact, the false conditions and interpretations of official scientific earth and natural history would be revealed in a flash.

Conclusion: Fossilisations and thus long–term preservation of bones can only take place under catastrophic conditions (e.g. volcanic eruptions, deluges) and/or if the bones are kept in an airtight place. On the other hand, not fossilised bones can also be preserved under extreme conditions such as freeze–drying or mummification (e.g. exposure to very high heat), if these conditions remain constant and lasting protection is provided against mechanical, biological or chemical decay. It is, however, unlikely that any type of preservation over millions of years can be achieved in this way. Therefore, if

Fig. 18: Time Impact. *An several metres thick layer (S) —scale (M): 2 metres — was formed in one day on by the air-fall debris from the last hours on May 18, 1980, and was then covered by several finely banded layers (G) on June 12, 1980, in only a few hours by hurricane-velocity surging flows from the crater of the volcano. Thick and/or numerous thin layers are not proof of long periods of time. This layer is overlain by a thinner massive mudflow deposit on March 19, 1982. The graduation and measurement of the earth's age (Lyell Dogma) based on geological layers as chronology, is an enormous error. Original photo by Steven A. Austin.*

in Africa human evolution is supposed to have been in a series of developments, like a range of car models, an ash—drift layer must have settled over the various rungs of the evolutionary ladder at certain intervals and preserved them. But is it correct that this supposedly gradual, imperceptibly slow evolution was accompanied throughout by a series of natural catastrophes over millions of years? Or was there really only a single, staggered catastrophe horizon, as would appear to be confirmed by finds of *Australopithecus* and modern humans in the same geological layers?

It is in fact the case that the total number of fossilised remains for the period from the Neanderthals through to *Australopithecus afarensis* (Lucy), i.e. for a period of almost four million years, can be spread out on a single billiard table. The Evolutionists answer to this question is: We will never have a fossil inventory representing even one thousandth of one percent of all the individuals who have ever lived (Tattersall, 1995).

With this argument, evolution research has carte blanche to arbitrarily interpret individual finds and thus manipulate them.

Until recently, it was believed that non-fossilised DNA aged between 2–1 million years was a biological impossibility. Rightly so! Even this period is far too long. How long do corpses buried in coffins, several metres below the earth's surface, remain preserved? For decades, centuries but at most, millennia. This book records a number of finds of not fossilised dinosaur bones and even DNA. In view of this not fossilised organic material, can one simply go back to normal and categorically assert that the animals from which these remains come lived 65 or more million years ago?

Paul Sereno's fossil finds, with aquatic dinosaurs and contemporary primeval crocodiles beside the fossilised bones of cows and humans in the same surface sand layer in the Sahara are apt witnesses to the fact that dinosaurs lived at the same time as modern humans, and that wasn't 65 million years ago.

If one cuts out the phantom periods of this shortened time horizon, Australopithecines lived – like Lucy – at the same time as dinosaurs in what was then a fertile area with large lakes in the region of today's Sahara. If one leaves the phantom periods in and insists dogmatically that the dinosaurs became extinct 65 million years ago, then one can draw the conclusion that contrary to evolution theory, primates lived at the same time as dinosaurs: this is exactly the opinion expressed by the researcher, Simon Tavare from the University of Southern California in *Nature* (Vol. 416, 18 Apr 2002, pp. 726–729).

To the horror of those who believe in evolution theory, Sudhir Kumar and Blair Hedges of Pennsylvania State University came to the conclusion, after studying fossil DNA, that most mammal species were already alive 100 million years ago and not – as posited by scientific opinion – only 65 million years ago after the extinction of the dinosaurs, as reported in *Nature* (Vol. 392, 30 Apr 1998, pp. 917–920). Large mammal species and dinosaurs apparently co-existed. Up to then, it was believed that only very primitive mammals no bigger than rats, constantly fleeing dinosaurs, lived during the earth's middle period. Only at night, when their enemies were asleep, did they dare to go out under cover of darkness to gather a simple meal of a few insects. Their survival strategy was "don't get noticed"! The victory march of the mammals supposedly began only after the extinction of the dinosaurs, when they filled the newly available biological niches. This dogma of evolution theory which has been upheld for almost 150 years collapses with the study results published in *Nature*. This is why the dogmatists had to contradict them: a heated debate about the results was carried on by scientists via internet on Nature.com.

There is, however, a growing body of proof for the coexistence of large mammals and dinosaurs. In the 1990s several expeditions led by the American Museum collected hundreds of mammals in Asia, which show that the diversification of mammals had already started in the Cretaceous, during the lifetime of the dinosaurs and not, as

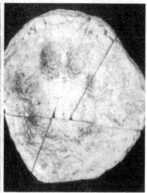

Fig. 19: Tracks. The fossilised tracks of large mammals in Cretaceous period layers, which also contain the footprints of dinosaurs, have been documented in several States of the USA. The photos show examples of the eastern edge of the Llano Uplift in Texas, which lay on the western coast of the Cretaceous inland sea in North America. Left picture: dog paw. Right picture: big cat, possibly saber-tooth tiger.

dogmatically propagated by the Evolutionists, afterwards (*Nature*, Vol. 398, 25 Mar 1999, p. 283; cf. Vol. 396, 3 Dec 1998, pp. 459–463). These studies however also show that mammals only appear relatively late in the early or middle Jura and not before in the Triassic, as was still being stated in 1997 in technical literature (Dingus/Rowe, 1997). In other words, the appearance of mammals took place 20–40 million years later than previously thought (*Nature*, Vol. 398, 25 Mar 1999, p. 283). They therefore only appeared at a time when dinosaurs completely ruled the world, and not before as had been proclaimed by science! Since the development of more highly evolved mammals was pushed back still further in time from the Tertiary to the later dinosaur era (Cretaceous), the dinosaurs thus lose their world ruler status as the species which filled all biological niches during earth's middle period.

The dogma of evolution theory relating to the sole ruler ship of the dinosaurs was finally destroyed in 2005 with an article published in *Nature* (Vol. 433, 13 Jan 2005, pp. 149–152).) Two new larger mammals were discovered, alongside other mammals (*Nature*, Vol. 421, 2003, pp. 807–814), in a 128–139 million year old formation from the early Cretaceous in the north–eastern Chinese province of Liaoning. One of the specimens (Repenomamus giganticus) was over one metre in length and weighed 12–14 kilograms. Based on his size and the appearance of the bite, scientists guess that he, unlike his relatives, was predominantly a carnivore. In theory, he could have been a scavenger but his long and pointed front teeth and the powerful chewing muscles suggest he was a predator. This animal did not, therefore, skulk secretly, but competed for prey and hunting territories with the dinosaurs. The established roles – mammal as prey and dinosaur as predator – are reversed.

Another mammal even ate a dinosaur! A smaller mammal, around 60 centimetres in length and weighing 4–6 kilograms (Repenomamus robustus) was first described scientifically in 2000 as a species of opossum. The new find was sensational in that with the almost complete skeleton, researchers were even able to analyse the stomach content. It contained a dinosaur – a young 14 centimetre long Psittacosaurus, which was more or less swallowed in one go. These herbivorous parrot dinosaurs grew to lengths of up to 2 metres (cf. Zillmer, 2002, p. 165).

Why is this find so well preserved? A complete animal with stomach contents only remains preserved as a fossil if it was buried airtight under catastrophic conditions, so that it could not decay. In the present case, it is thought that a volcanic eruption led to a mass extinction of dinosaurs, mammals and amphibians, who were then enclosed in the tufa layers of the Yixian Formation (*Nature*, Vol. 421, 2003, pp. 807–814). This Formation has long been the site of *spectacular fossil finds. The birds there were, like in the Messel Pit in Germany, probably killed by volcanic gas, primarily carbon dioxide and fell into the volcanic ash which hardened shortly afterwards; on the other hand, birds with broken wings also fell into the lake after the explosion (cf. BdW, 8/2002, pp. 54–63).*

It appears that in this case too, a time impact took place, because the corresponding 125 million year old layers of the Zixian Formation from the early Cretaceous (*Nature*, Vol. 401, pp. 262–266) were formed suddenly after a volcanic eruption. This is why the fossils contained therein represent only a snapshot scene of life: this layer was formed very fast in a matter of hours or, at most, days. No evolutionary development can be deduced from this, even if there were several such volcanic eruptions. If we scrap the Tertiary, as discussed earlier, dinosaurs and larger mammals coexisted a few thousand years ago, until before the Deluge. Are the radiometric determinations of volcanic stone wrong?

3 Dating and Arbitrariness

An astonishing rejuvenation: in 1996 researchers claimed that artefacts found at the Jinmium Cliffs in Queensland, Australia, could be 176,000 years old. According to a report in "Nature" (Vol. 393, 28 May 1998, pp. 358–362), an examination of the artefacts using new techniques showed them to be less than 10,000 years old. The moral is that every method of testing achieves different results, none of which correspond with reality.

Hot Spot Volcanism

Given natural catastrophes in historical time, the conditions for fast fossilisation processes exist. This time–impact scenario introduced in *Mistake Earth Science* has been confirmed under laboratory conditions: American scientists succeeded in fossilising wood in only a few days. As such the natural process which would take Nature millions of years, was achieved in a laboratory. In a short space of time the wood's organic material was gradually replaced by minerals – such as crystallised silicon dioxide – so that the original structure remained completely intact. As in nature, this process had to be carried out in airtight conditions, because if the wood is in contact with air, it cannot fossilise pore for pore over millions of years, as it would rot first (*Advanced Materials*, Vol. 17, January 2005, pp. 73–77). In nature, trees only fossilise if they have, for example, been buried under airtight lava layers. For this to happen, natural catastrophes must occur to ensure the necessary pressure and heat.

In East Africa, apparently, the most fossilised members from the chain of human evolution are found. So we have to look there for catastrophes responsible for the preservation of these fossils. The East African Rift System (Rift Valley), known in geology as the Rift Zone or Rift (a narrow crack or fissure in the earth's crust), is in East Africa. The Rift Valley in East Africa is part of the *East African Rift Zone*, which belongs to a rift zone which extends from Syria, through Lebanon, the Valley of Jordan and the Red Sea as far as Mozambique in southern Africa.

In the region of the East African Rift there are two raised domes, the Kenya Dome and the Afar Dome. Both are known as mantle plume hot spots. More recent geophysical opinion is that massive streams of hot magma form in the boundary layer between the outer earth crust and the lower earth mantle at a depth of 2,900 km and rise up as far as the earth's crust in the form of a small pillar – so called plumes – and under the lithosphere (earth's crust and upper part of the mantle) in a mushroom shape, forming a type of dome which softens the brittle earth's crust from below and can break through at weak points. The hot material then burns its way through the earth's crust like a cutting torch, whereby volcanoes form far from the edges of the tectonic plates. They are described as hot spots. This in turn leads to hot spot volcanism, through which large areas can be covered with lava which hardens into dense basalt layers. At the same time, however, the ash-drift expelled during the volcanic eruptions is scattered over wide areas, such as in East Africa, and this occurs several times, extending over greater or lesser areas depending on the intensity, through irregularly distributed volcanoes and breaks in the earth.

In this way, several geological layers may form in a very short space of time which may, however, also last years or decades. This would explain the formation of layers of ash–drift containing the ape and human fossils. The supposedly 3.6 million year old Laetoli footprints, which are ascribed to modern humans, were also left in an ash-drift layer.

Furthermore, mudslides and floods are caused by volcanic eruptions. During the 1980 and 1983 eruptions of Mount St. Helens in the US State of Washington for example, earth layers of up to 50 metres thick were formed in only a few hours.

If objects, animals or humans are enclosed in these sedimentary layers, they may be lying up to 50 metres below the earth's new surface. If these sediments contain a hardener (plaster, lime, calcite cement or grit cement) and sufficient water is available, these layers will harden within only a few hours into limestone or sandstone layers. Whether these sediment layers are loosely or densely packed or perhaps even comprise solid rock, depends on the quantity of hardening agent (hydraulic binding agent, minerals) contained therein and on the proportion of hardener to water (= water-cement ratio). In this way, sediment and rock layers form very fast during natural catastrophes. I described this scenario which I first introduced in *Darwin's Mistake* as "Natural Concrete Theory" to the horror of geologists and geophysicists, because it exposed the grain for grain slow growth of the earth's layers as a great 150-year old bluff. At the same time, the age of the earth reduces many times, to the power of several tens, so to speak: several zeroes should be knocked off the established age of the geological layers. A good illustration of this is my assertion with regard to the Grand Canyon (see p. 44ff.).

With regard to the observations made here, it is important to now ask when this intense volcanism which preserved ape and human fossils, took place.

The Grand Canyon's Only Young

Is the relatively high frequency of human fossils in East Africa due only to the effect of volcanic eruptions? Were humans witness to the tearing apart of Africa, when the Rift Valley formed? Everywhere along the line, indigenous peoples have traditions of great changes in the structure of the land. This opinion is underlined by geological phenomena, as some of the rift slopes are so bare and sharp, that they must be more recent having formed during the lifetime of humans (Gregory, 1920, p. 31ff.). Did these tremendous geodynamic processes therefore take place only a relatively short time ago?

In 1920, Gregory shared the opinion of the famous Vienna geologist, Prof. Eduard Suess: The East African Rift System is connected with the rising of the mountain chains in Europe, Asia and on the American continent. Thus the time of the last uplift, if established, would also show when Africa experienced this great split.

R.F. Flint (1947, p. 523) wrote: The earth was under pressure and its crust burst along a meridian almost the entire length of Africa. The mountain chains on the floor of the Atlantic could have been formed by the same cause; and the time of the breach and the folding must coincide with the period of mountain formation in Europe and Asia. These mountains achieved their present height during the lifetime of humans; the East African Rift System was largely formed during Man's lifetime, at the end of the Ice Age. Did humans witness the formation of the mountains, the middle–oceanic ridge and the East African Rift System?

The tension responsible for tearing open the East African Rift (and the further rifts in the Red Sea and the Valley of Jordan), was caused by the earth's expansion as already described (cf. Pascual Jordan, 1966, p. 24ff.; Zillmer, 2001, p. 98ff.). If one takes the *true polar wander* (movement) into consideration, the East African Rift lies, in the view of P. A. Vening Meinesz (1943) exactly along the same line as the break in the earth's crust, and also in the same direction as the north to south chain of volcanoes on America's west coast. This fracture zone in western America is still moving and is of interest to us because it is there that, as in East Africa, volcanic eruptions have preserved controversial finds relating to human history.

During the expansion of the earth's volume, the earth's crust tore apart and the network of cracks creating the hitherto non–existent tectonic plates, now separated from each other by tears (middle oceanic ridge) and deep–sea rifts (as before the western coast of South America). Modern geophysicists come to false conclusions here because they only believe satellite calculations and complicated computer simulations which, however, only work on the basis of the scenarios and data fed into them. When interpreting geological phenomena, today's earth scientists do not consider the strength of materials to the laws of which the earth's crust is also subject, as already pointed out by Gabriel Auguste Daubrée (1880) and Ott Christoph Hilgenberg (1949) after their respective experiments.

With the enlargement of the globe, gravity also changed on earth. Large forms, such as the sauropods which reached lengths of up to 50 metres, were no longer able to survive, as under the given conditions only animals up the size of an elephant could live, but not larger land animals such as giant dinosaurs! Did the scenarios of earth's expansion and the folding of the mountains take place during the lifetime of Man? The legends of the

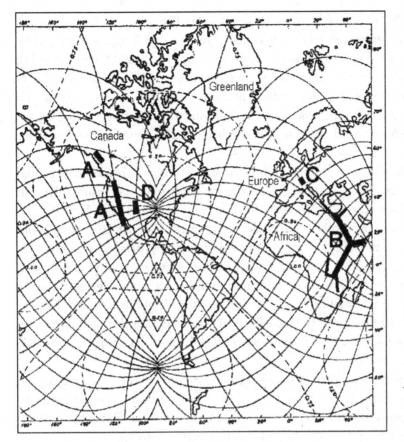

Fig. 20: Fracture Lines. The drawing shows the fracture network of the sialic earth crust (= silicon-aluminium), formed through poler movement of 70° along the meridian 90° east, according to Vening Meinesz (1943). Taking the theory of the strength of materials into consideration for the interpretation of geological events, we have a fracture network of the earth, which can, on the one hand serve as proof for Wegener's shifting theory (1941) but can equally support the theories of Keindl (1934) and Hilgenberg (1933) who claimed that the earth (cf. Zillmer, 2002, pp.71-124) increases in volume with increasing time. It is interesting that the rift zones (riftogenes) after Nikolaev (et al. 1984) all lie in the direction of the fracture lines: Basin and Range Province on America's west coast (A), the East African Rift Valley system and the Red Sea (B), the Rhein-Rhône Rift (C) and the Rio Grande Rift (D). The rift zones A and B still represent active volcanic areas (mantle plume hot spots). And it is precisely here, as well as buried and preserved under layers of lava and ash, that controversial finds in the Sierra Nevada, as well as ape and human skeletons in East Africa have been made.

indigenous peoples of North America in any case report of the formation of the Grand Canyon and, in South America, of the rising of the Andes. In India, a "rapid towering up of the Himalayas, the Tibetan Highlands, Pamir and other mountain chains" is mentioned (Allchin/Allchin, 1982, p. 14), whereby even settlements of the younger Old Stone Age were buried under layers of clay (cf. Heinsohn, 1991/2003, p. 54).

That East African myths speak of the formation of the *East African Rift* should be seen in the same light of global transformation. What do the geologists believe? The East African Rift is said to be 20 million years old. Because of its great age, the development of humans supposedly only took place east of the Rift. In 2001, the paleo-anthropological community was shocked when on 19 July of that year, the team led by the French anthropologist, Michel Brunet (*Nature*, Vol. 418, 11 July 2002, pp. 145–151), found the skull of a primeval human in Chad, which was called Toumai and apparently displayed some characteristics of *Homo erectus*. The find was located to the west of the East African Rift, i.e. on the wrong side, because in contrast to the eastern-lying region where Lucy once toddled around, the Toumai site belonged to "ape country" and not "human country".

In the meantime, the skull was examined in Europe. Since then it has been clear that the history of human evolution in East Africa cannot be correct, as Toumai lived in Central Africa 6–7 million years ago, in other words around 3 million years before Lucy. Meanwhile, however, other researchers are contesting that this Toumai individual could have walked upright – although they have not undertaken any detailed examinations (*Nature*, Vol. 419, 10 Oct 2002, pp. 581–582). Michel Brunet's response was that *Australopithecus afarensis*(Lucy) had also been interpreted as a young gorilla by her detractors and that the same classification of his find as a female gorilla would not fit the facts (*Nature*, Vol. 419, 10 Oct 2002, p. 582). The Harvard professor, Daniel Lieberman, described this find from Chad as a "small atomic bomb". Does the history of human evolution have to be rewritten?

The whole problem is caused by the propagated notion of very long geological periods. Because if the *East African Rift* is far younger than supposed, this problem no longer exists and the western–lying "ape country" can also become "human country". Everything depends on dating. So let's take a closer look at dating methods.

Incalculable Lava

Radiometric dating methods were first developed in the 20th Century. These new radiometric measuring methods did not lead to any change in the basic geological assumptions, which were already established, completely arbitrarily, in the middle of the 19th Century. It was in fact, the speed of evolution which was used as a yardstick, because the supposed slowness of the processes involved required very long periods of time, in order to give the game of genetic chance sufficient time.

In spite of all the new techniques available, there is as yet no method for absolutely dating sediment layers from limestone or sandstone. However, it is only in these sediments that fossils are found, unlike with granite and basalt which contain no fossils. For some years, it has been believed that these volcanic stones can be dated absolutely and to do this, potassium–argon testing is usually the preferred method.

All dating methods, however, are based on certain assumptions. If these are not given or are only partial, the dating results may be incorrect or even irrelevant.

With the potassium–argon method there is a basic problem with the presence of superfluous argon. The radioactive decay of the unstable (radioactive) isotope of potassium-40 and/or the resulting argon-40 during the cooling of volcanic stone is measured. To achieve a precise measurement, all of the argon present at the outset would have to be dispersed from the minerals, so that the starting point would be zero – an absolutely essential requirement. However, it is not known how much argon was present when the cooling in the minerals started and/or how much argon comes from the atmosphere. The existing argon cannot be distinguished from the radioactive product of decay, argon-40 which forms during the cooling.

Using this method, it is calculated that the more argon the stone contains, the older it is. The ages obtained with this method are often also described as model ages or, in this case, potassium–argon ages. Another important prerequisite for this method is that no further perturbations may take place during the radioactive decay of potassium-40 during the course of this lengthy period of time.

Stone however has the property of absorbing argon. Because of this unavoidable enrichment, a greater quantity of argon is measured than can be present exclusively through the radioactive decay of the original quantity of potassium-40. Under catastrophic conditions a radical reduction of argon-40 to almost zero is also possible, because even under normal conditions, argon can penetrate surrounding water or stone or, because it is water-soluble, be washed away by microcirculation.

If the effects of each of these influences, potentially over millions of years, were evaluated, this would lead to errors in calculation of unknown dimensions. It should also be clarified as to whether, under normal conditions, a constant rate of decay can be accelerated or decelerated under certain conditions or circumstances. Since neither the proportion of argon in stone at the start of cooling nor the quantity of argon absorption or dispersal are known during the period being measured, age dating with this method must be subject to error. The simple question is: how big are the errors? Or are the age datings entirely wrong? Where volcanic stone of known age, formed in the last 200 years, has been tested, the potassium–argon age has been given as up to 200 million years old – too old by a factor of one million.

In order to prove that radiometric measuring methods can produce correct results, a series of tests needs to be carried out. It seems obvious that the test material for dating should be lava from a volcanic eruption, the time of which is exactly known.

Let's take a look at my favourite volcano, Mount St. Helens. The present lava dome began to form after the last eruption on 17 October 1980. Extremely viscous dacite lava swelled out of the volcano crater, like toothpaste coming out of a tube, and formed a mountain-like dome.

Since the age of the lava in this dome is known, the dating of radioisotopes should produce a fairly young age. In June 1992 Dr. Steven A. Austin (1996, pp. 335–343) extracted a 7.5 kg block of dacite from high up in the lava dome. Part of this sample was crushed, sieved and processed into stone powder and four mineral concentrates, including silicate compounds (amphibole, pyroxene). These were then handed over to the Geochron Laboratory in Cambridge, Massachusetts for radioisotope dating. The laboratory was not told that the samples came from the lava dome on Mount St. Helens and were, at that time, only 10 years old.

Radiometric dating using the potassium–argon method produced the following ages for the individual minerals from the same sample:

- Stone powder $350,000 \pm 50,000$ years
- Feldspar compound $340,000 \pm 60,000$ years
- Amphibole compound $900,000 \pm 200,000$ years
- Pyroxene compound 1 $1,700,000 \pm 300,000$ years
- Pyroxene compound 2 $2,800,000 \pm 600,000$ years

Firstly, the result allows for a very large margin of error but what is even more interesting is the great age! These were samples which were only 10 years old. In other words, appropriately high levels of argon had been found in the samples; but as a prerequisite for the accuracy of the dating, the radioisotopic stopwatch should have been set to zero prior to the eruption and only started ticking when the eruption took place. Since the ages of the samples have been set far too high, there was absolutely no firm basis for a correct reading. This may be compared with the incorrect timing of a runner who gets off to a flying start whilst those he is running against have to begin from a resting state.

The extremely broad spread of ages for the various minerals, all of which come from the same sample, is also instructive. Conclusion: the ages measured using the same method are largely dependent on the type of material tested. That instead of 10 years, the dacite was dated to such incredible figures ranging from 340,000 to 2.8 million years shows that corresponding readings of volcanic stone of unknown age cannot generally produce accurate results.

Does the argon content of the dacite from Mount St. Helens lie outside the measurable scale, as claimed by orthodox scientists (Ries, 2003)? If any value at all for the argon content was found, then the reading cannot be outside the measurement scale; in fact it is just the opposite: because something can be measured, the situation is therefore reversed. Thus the original criticism of dating methods is confirmed, because nobody knows how much argon–40 was present in the lava at the outset. This is why age datings with the potassium–argon method are not meaningful, as it is not known whether the argon stopwatch in old lava stood at zero – because nobody measured it at the time. This is an area which calls for belief rather than knowledge.

If I assert that the volcanic stone of the Andes was formed 5,000 years ago and the radioisotope stopwatch was not set to zero at this time but had a high starting value, nobody can contradict me. In other words: if the mountains (including the Andes and Himalayas) rose only a few thousand or tens of thousands of years ago, long-term dating is not possible and the readings will produce a fantastical age which is far too high. Or to put it in another way: if the lava dome on Mount St. Helens had been formed not 10 but 5,000 years ago, i.e. before history was being written, the potassium-argon readings would have been believed and the geology books would say that the age of 2 million years had been proven in tests.

Completely Wrong Measurements

Is the dating of the Mount St. Helens lava dome an isolated case? Let us first analyse the dating of an event, whose date one believes is known (provided that the official historical records are correct): the eruption of Vesuvius in 79 A.D., which buried Pompeii. The dating resulted in an age which was almost twice as high: 3,300 ± 500 years (Renne et al., 1997). Officially it was claimed that this value was acceptable, even though there was a discrepancy of about 70% (*Science*, Vol. 277, 29 Aug 1997, pp. 1279–1280).

As the sample seemingly contains too much argon, the age reading was mathematically adjusted to fit the real age. That is why it is claimed that historical events can be dated to within a maximum margin of error of 5% using this method. If one knows what the result is supposed to be, it's easy to adjust the readings. In addition, the dating results vary widely depending on the isotopes and type of mineral examined.

The 1997–1998 research report from the Institute of Mineralogy of the *University of Münster* (Germany) dated 5 November 1999 stated that age datings of high–pressure metamorphites in the Saih-Hatat Region in Oman, using a combination of two methods – argon-argon (Santa Barbara) and rubidium–strontium (Münster), produced significant differences in age, since these high-pressure stones display an excess of argon.

Are these isolated cases? No: G.B. Dalrymple (1969) has listed several age datings of lava samples from volcanic eruptions in recent history, obtained using the potassium–argon method (*Earth and Planetary Science Letters*, Vol. 6, 1969, pp. 47–55): see Fig. 21.

Since it is known that age datings produce false results, certain ratios of different isotopes are combined in two different ways (three-isotope diagrams), and the half life of the isotopes are then used to calculate the common onset of their decay. If similar results are achieved, it is assumed that the radioactive decay proceeded steadily without perturbation. The result therefore depends entirely on assumptions and the expected result. The same goes for the claim that the absorbed part of the argon–40 can be measured by also measuring the argon-39 content (Stan/Hess, 1990, Werner et al., 1997) or the argon-36 content (Dalrymple, 1969; Lippolt et al., 1990). The ratio of these two isotopes in the atmosphere is deemed to be constant over hundreds of millions of years although, for example, at the outset a large quantity of one or more isotopes can be simultaneously absorbed.

volcano eruption	year	lowest age-dated	error in years
Hualalai Basalt, Hawaii	1800–1801	1,330,000	1,328,000
Mount Etna Basalt	122 BC	170,000	168,000
Mount Etna Basalt	1792	210,000	210,000
Sunset Crater Basalt	1065–1065	100,000	99,000
Mt. Lassen Plagioclase	1915	80,000	80,000

*Fig. 21: **Too old.** Known volcanic eruptions dated using the calium-argon method resulted in fantastical ages. After Dalrymple, 1969.*

In *Science* (Vol. 277, 29 Aug 1997, p. 1280) it is said that: *Whether the initial $^{40}Ar/^{36}Ar$ ratio reflects a magmatic argon isotopic composition or is biased by argon adsorbed after the eruption cannot be determined from our data.* For non-specialised users, however, the internet lexicon, Wikipedia (German text) claimed as late as January 2005 that: "in the case of the potassium–argon system, the argon–40/argon–40 measuring technique is a particularly sophisticated one which rules out falsifications".

It must be underlined once again that the age of stones cannot be measured directly, but that the number of isotopes are counted and a calculation based on today's rates of decay then establishes a date in the past. However, these methods lack any form of calibration against known objects: the absence of data on calibration curves, standard deviation and background emission is replaced by assumptions, e.g. that there is no dispersal of the child isotope or that no changes could have occurred through adsorption (*Lexikon der Physik*, 1999, Vol. 3, p. 199). One must also consider that both the number and ratios of the isotopes are influenced by later metamorphic transformation of the stone.

The period of time in which half of the radioactive isotope decays (half–life time) has previously been considered as a laboratory–tested "calibrated" clock. This dogma has been gently shaken: in *Science* (Vol. 282, 4 Dec 1998, pp. 1840–1841) an error model was published which showed the variations in half–life times in various radioactive isotopes based on theoretical and experimental assumptions. Conclusion: nuclear decay is not an exact science.

When potassium-40 decays, it is not only argon–40 which is formed, but also calcium-40. This calcium forming through radioactive decay is not officially recognised, as it cannot be differentiated from normal calcium which is often present. However, calcium is a silvery-white, soft metal which reacts strongly with water. It is the basic element in the formation of limestone (calcium carbonate $CaCO_3$) while cement and chalk are important calcium compounds. In water, calcium carbonate is present in dissolved form, i.e. ions. In the event of physical changes to the water (evaporation, heating, swirling), the calcium carbonate falls out. In other words if, for example, because of catastrophic events (meteorite impact) and the consequent onset of a greenhouse effect, the water was simultaneously heated and swirled, lime, sand and chalk layers would form relatively quickly. Would this explain the widespread formation of the previously mentioned chalk cliffs (inter alia in the Baltic and North Seas)?

The deluge model presented in *Darwin's Mistake* offers the additional explanation that after a meteorite impact, not only is new limestone formed, but with the release of heat (Greenhouse Effect) two parts water per molecule of $CaCO_3$ (calcium carbonate) are also formed. In other words: after an asteroid impact, new limestone is formed and the sea level rises, a phenomenon which has previously been ascribed to the melting of the "Ice Age" glaciers (Zillmer, 1998, p. 80ff.). My "natural concrete theory" describes the rapid formation of various sedimentary layers which hardened quickly. However, the otherwise mysterious formation of calcium through the previously mentioned radioactive decay of potassium-40 gives further food for thought: Are intense volcanic eruptions caused by the existence of vast quantities of calcium and the, in terms of earth's history, relatively late formation of limestone? Did this process take place during the earth's expansion?

Too Few Noble Gases

The transformation of uranium and thorium nuclei into radioactive isotopes occurs by means of alpha and beta decay, usually with the subsequent emission of gamma radiation. The alpha radiation consists of so–called alpha particles, in other words out of helium nuclei, formed from two protons and two neutrons. Because of their unusually high binding energy (28.3 MeV) the helium nuclei are extraordinarily stable. Helium has the lowest melting and boiling point of all known gases and cannot be frozen out by atmospheric pressure – making it extremely resistant.

Solar corpuscular radiation, known as solar wind, comes from the sun's corona. This is made up of both hydrogen and helium nuclei. If we imagine for a moment that no helium nuclei from the solar wind remain in the earth's atmosphere, then according to conventional science, helium can occur exclusively as alpha particles through release during radioactive decay with alpha radiation.

If we assume, on the basis of radiometric datings, that our oldest rocks have been undergoing radioactive decay for a period of up to more than 4 billion years, there must be a huge number of helium nuclei in our atmosphere. Using sensors, the emissions of helium nuclei from the earth's crust into the atmosphere have been measured. If one assumes that all of the helium remains in the earth's atmosphere, a mere 10,000 years would be sufficient for our atmosphere to achieve its present helium level.

Is it possible that these very light helium nuclei can be passed into space, so that there is only a constant exchange? According to Thomas Gold, former professor at *Harvard University*: Chemically reactive elements would have had to have been removed from the atmosphere through some sort of chemical process; but the noble gases, which are both heavy enough to be retained by gravity, and chemically inert so that they cannot have been absorbed into the earth's crust – must still be present (Gold, 1988, p. 36). But even if one assumes that a high volume of gas were emitted into space, the age of the atmosphere can still only be calculated as a few hundred thousand years at most.

The real age is significantly lower, because the maximum energy which is stored in the earth's magnetic field can be calculated. This rate of energy is reached after only 8,700 years in the event of free radioactive decay in the earth's crust. This age is only slightly higher than the period required for an earth-bound, loss-free enrichment of the atmosphere with helium.

Other scenarios also indicate that our atmosphere is young. If one calculates a loss rate of 15% in the strength of our magnetic field which was measured in the last 160 years, and transpose this rate onto the earth's past, the strength of the magnetic field 22,000 years ago would have been so powerful that no life form could have survived (discussed in detail in *Mistake Earth Science*, p. 301ff.).

However, in no event should this argumentation be taken to infer that the planet Earth is young, but rather that the earth's crust was almost completely transformed with a new or reshaped atmosphere, a few thousand years ago. At this time, radioactive decay accelerated significantly, resulting in the formation of calcium and water, and a stress peak was created.

At the same time, cosmic radiation of the earth increased as well. In the upper layers of the atmosphere, almost all ultraviolet–C radiation and at least 97% of ultraviolet–B radiation is absorbed by molecular oxygen. This leads to the formation of ozone, provided that the atmosphere contains enough oxygen. The ozone layer works, in turn, as a filter against ultraviolet rays. If the oxygen level in the atmosphere or the ozone level in the stratosphere reduces, ultraviolet–B radiation in particular increases, and these rays can penetrate through to the earth. The more energy the radiation contains the more damaging its effect on bio–mass: ultraviolet–B radiation can cause abnormal changes to the cells of all living creatures, it is instrumental in the formation of cancers, it weakens the immune system, increases the rate of deformity, causes genetic damages and can, if sufficiently intense, lead to death. New research has shown that animals, and probably humans as well, become infertile if exposed to high levels of ultraviolet-B radiation.

A dramatic increase of (infertile short–wave) ultraviolet–B radiation (cf. Zillmer, 1998, p. 154) and radioactive radiation in conjunction with an increase in gravitational force during the Deluge, was probably the case for the mass death of 80% of all species (Zillmer, 2001, p. 286ff.). Amongst the survivors of this scenario, a diminution of the once massive fauna took place, giving rise to the small plants, animals and humans who now populate our planet.

My argumentation was recently confirmed: it is now believed that increased radioactive radiation was responsible for mass extinction 443 million years ago, causing the extinction of the trilobites. The ozone layer was destroyed and the intensity of the UV-radiation increased (*New Scientist*, 27 Sept 2003, p. 17). The previous theory ascribed the extinction of erstwhile aquatic animals such as starfish, sear urchins, cuttlefish and arthropods to the onset of the Ice Age. But nobody has yet been able to explain why an Ice Age should suddenly begin in the middle of a period with a very warm climate.

Conventional science's view of how the earth was formed and the processes going on inside it is also, contrary to the claims of the geophysicists, fragmentary. The report which stated that within the earth helium atoms are apparently also released without radioactive decay (*Earth and Planetary Science Letters*, Vol. 31, 1976, p. 369–385) came as a sensation. If one takes these free helium atoms apart from radioactive decay, then the age of the atmosphere must be reduced still further. It is time for the blinkers of old dogma to be torn away from our minds and from research. It is high time for a paradigm shift.

4 Pack of Lies: Descent of Man

The world of science uses a knowledge filter, which sieves out unwelcome material. This knowledge filtering has been going on since the end of the 19th Century and continues today. Finds which contradict conventional wisdom are rejected without any serious examination of the object in question. Once the rumour that a particular find is dubious has done the rounds in the scientific community, this is usually enough reason for most scientists to give the find no further thought. A cloak of silence is then thrown over the find. Younger scientists thus learn nothing about the existence of finds which are controversial, or which even turn conventional wisdom on its head; they believe, indeed, they are absolutely convinced that they are being soundly and omnisciently trained by the university system. This is why earlier descriptions of controversial finds must be preserved for a time when a scientific revision can take place – a sort of Galileo case for the scientific world – until such time as an unprejudiced, critical generation of researchers has grown up, which bases its theories on facts and not vice versa.

Knowledge Filter

In October 1998, Fritz Poppenberg's movie "Is the Bible Right After All? The Theory of Evolution Lacks Evidence" was broadcast by the German television channel *Sender Freies Berlin,* whereupon three scientists made official complaints. The documentary was marked with a lock flag and "may no longer be scheduled for viewing on television" (Kutschera, 2004, p. 248). When, during a speech with the title "Evolution: the general theme of biosciences" given at the annual conference of the *Society of German Biologists* on 27 October 2002, Prof. Dr. Ulrich Kutschera of the *University of Kassel* warned explicitly against Poppenberg's movie, the books *Darwin's Mistake* (1998) and *Ein kritisches Lehrbuch* (Junker/Scherer, 2001), the "Arbeitsgemeinschaft Evolutionsbiologie" (Working Group for Evolutionary Biology) was established after the conference, to prevent any further anti-Darwinian influence on education and the public and to secure the jobs of the evolution theorists.

When we speak of the suppression of evidence against evolution theory, it is not a case of individual scientific conspirators who wish to deceive the public. No, it is much more a continual process of systematic knowledge filtering, which appears harmless, but over the course of time has taken on significant dimensions and is still growing, practically becoming impenetrable for undesirable information.

As such, the controversial themes I raised in *Darwin's Mistake* were also blacklisted. A multi–award winning director wanted to produce an entire series for the public television companies but was told in no uncertain terms that anybody making even a single documentary with Zillmer could forget about getting any more commissions. To give another example, two conservative assessors have the power to decide on what research findings are published in the journal *Science*. Thus controversial research is not seen by other scientists and is most certainly not available to the public. The rejected research reports are then published in journals, which are read worldwide by about 500 specialists. The result is that controversial research is buried within individual specialised disciplines. Scientists who present and discuss explosive evidence are denounced as unsound and their professional careers are impeded or even suspended.

What makes it worse is that today every expert in fairly closely related disciplines is merely a member of the generally-educated public, who without a comprehensible and good introduction cannot engage with the associated problems so that a full understanding which depends on this, is no longer possible. As such, at most only a handful of experts per discipline have a flawless monopoly in their grip. Nobody is authorised to discuss research findings relating to their particular discipline, because anybody else, including professors of closely related disciplines are treated as the initiated, i.e. ignorant, who supposedly have no understanding of the material.

Fortunately for humanity, the internet has established itself as a source of information, so that information can be disseminated immediately. The concealment of explosive information cannot therefore function in the same way as before. It may be observed, however, that interested circles also use the internet to make denunciations, by using the discussion groups of educated scientists and drilled laymen to prop up old dogmas and to standardise the formation of opinions. A boorish tone which includes personal attacks is used and this discourages those who have a general interest from taking part in these discussion groups. Thus the goal is achieved; the flow of information is stemmed.

As has always been the case, the book is a source of information which has a longer lifespan and whose information content cannot be so easily wiped out. The present book aims to present to those who are interested in Man's origins and beginnings, various pointers, texts and material which are absent from today's standard textbooks and are not easy to obtain. The reader will see that the currently prevailing opinion on Man's origins requires fundamental revision.

Scientific Manipulation

Between 1951 and 1955, the anthropologist Thomas E. Lee from the *National Museum of Canada* discovered stone tools in "Ice Age" deposits at Sheguiandah on Manitoulin Island, the world's biggest lake island, located in Lake Huron in North America (Lee, 1966). These stone tools showed signs of advanced workmanship and, after several examinations by the geologist John Sanford from *Wayne State University* from 1952–1957, were said to be at least 65,000 years old and possibly as much as 125,000 years old.

Perhaps the best approval of these unsorted deposits as "Ice Age" deposits was the visit of some 40 or 50 geologists to the site in 1954 during the annual field trip of the *Michigan Basin Geological Society*. The sediments were presented to this group and there was no expressed dissension from the original explanation (Sanford, 1971, p. 7).

Opinions were split on the question of dates only. From the range of up to 100,000 years, it was agreed that the minimum age should be 30,000 years. According to the time ladder of human history, only Neanderthals or early–modern Aurignacian Humans could have fashioned these tools in North America.

The tools were not found on the surface but were excavated from several geological layers: The stratigraphic sequence of the sediments and the artefacts contained in each layer are definitive and without any doubt. Thorough excavation and examination of the sediments and the retrieved artefacts leave no room for doubt with regard to the undisturbed sequence of existing cultural layers (stratigraphy) (Sanford, 1971, p. 14).

New excavations have been undertaken in recent years under the guidance of the anthropologist Prof. Dr. Patrick Julig of the *Laurentian University* in Sudbury. According to his findings, the excavation sites were inhabited by indigenous people *before and after a cataclysmic flooding about 9,600 years ago* (Julig, 2002). Organic material in the corresponding layers of the turf moors was dated using the radiocarbon method. Having denied it in 1979 (Griffin, 1979, p. 43ff.), the anthropologist James Griffin of the University of Michigan was forced to admit in 1983 that these finds were genuine. However, this admission created the impression that the tools were only found on or near the surface of the turf moors. More recent site sketches (Julig, 2002) expressly mentioned that no tools were found in the glacier rubble which is considerably older than the moors.

The older original reports however offer no arguments against this mud flood hypothesis. According to Lee, the artefacts which were excavated from the layers of the turf moor and recognised as genuine, belong to an upper level with projectile tips, which lie above the "Ice Age" rubble layers. This younger stone tool culture is "Indian" in the broadest sense and was common across a large part of central Canada. In the level below, i.e. in older layers, completely different types of tools to those found on the turf moor layers close to the surface were discovered in the glacier rubble, contrary to Julig's site sketch (Lee, 1983).

Using conventional dating, the artefacts from the glacial rubble are far too old. This is why the finds of ancient stone tools from the lower–lying, thus older, layers had to be literally crushed beneath the wheel of science. As the artefacts are indisputably authentic, the attempt has been made to classify the stone tools as "post Ice Age".

With regard to this case, the new interpretation of the tool–bearing layers as "flood layers" is interesting and comes close to my interpretation (see Chapt. 2) although the age needs to be significantly reduced again, as in my view, even the older Ice Age rubble layers of the "Wisconsin Ice Age" can be seen within a post-Deluge time horizon. This would mean that the artefacts may be even less than 5,000 years old.

What's really interesting, however, is the light that this case sheds on the scientific community's practice of shutting out colleagues who do not submit to the teachings of the scientific authorities. Thomas E. Lee, the discoverer of the find sites, "was hounded from his Civil Service position into prolonged unemployment; publication outlets were cut off; the evidence was misrepresented by several prominent authors ...; the tons of artefacts vanished into storage bins of the National Museum of Canada; for refusing to fire the discoverer, the Director of the National Museum, who had proposed having a monograph on the site published, was himself fired and driven into exile;...and the site has been turned into a tourist centre ... Sheguiandah would have forced embarrassing admissions that the Brahmins did not know everything. It would have forced the rewriting of almost every book in the business. It had to be killed. It was killed" (Lee, 1966, 18f., cf. Cremo/Thompson, 1993).

Other similar finds have also been rejected by most archaeologists because they contradict this theory. In the 1960s at Hueyatlaco, about 120 km southeast of Mexico City, the geologist Virgina Steen-McIntyre, together with other members of a research team of the U.S. Geological Survey, uncovered high quality stone tools. The geology team which was funded by the National Science Foundation used several dating methods and put the age of the upper geological layer at least 245,000 years old.

After centuries of archaeological research in the Old and New Worlds our knowledge of early human history is so inexact that we suddenly realise: everything we've ever thought is wrong. Otherwise, the more detailed the collected geological data is, the more difficult it becomes to explain how several different and independent dating methods can have led to errors of the same dimensions (*Denver Post*, 13 Nov 1973).

Accordingly it was quite difficult for Virginia Steen-McIntyre to have her finds published at all. The printing of her report was postponed again and again until it finally appeared in 1981 in Quaternary Research, after she had expressed her discontent in a letter dated 30 March 1981:

The problem as she saw it is much bigger than Hueyatlaco. It concerns the manipulation of scientific thought through the suppression of "Enigmatic Data", data that challenges the prevailing mode of thinking. Hueyatlaco certainly does that! Not being an anthropologist, she didn't realize the full significance of there dates back in 1973, nor how deeply woven into our thought the current theory of human evolution had become. There work at Hueyatlaco has been rejected by most archaeologists because it contradicts that theory, period. Their reasoning is circular. *Homo sapiens sapiens* (H.s.s.) evolved ca. 30,000–50,000 years ago in Eurasia. Therefore any H.s.s. tools 250,000 years old found in Mexico are impossible because H.s.s. evolved ca. 30,000 years ago ... etc. Such thinking makes for self–satisfied archaeologists but lousy science!

Photographers are only authorised to print pictures of the Hueyatlaco artefacts if the "crazy date" of 250,000 is not mentioned and the age is stated as 30,000 years – a time horizon which can just about be aligned with the appearance of modern Man. Conclusion: for a good century, a knowledge filter has been used in science to get rid of undesirable material. Let's look at some other finds which don't fit into the concepts of evolution theory and geology.

Not Allowed

It happened on a June day in 1977. The German globetrotter, Ernst Hoening, found a tool fashioned by human hand, quite by chance in the 11 metre-deep Bison Gulch in northern Canada. Looking for further finds, Hoening unearthed fossilised bones of humans and animals, as well as hand scrapers and hand axes from distant prehistory. In addition, a footprint in a geologically-dated stone sheet was discovered, lying directly below the boulder clay of the last Wisconsin glaciation. The geological dating of the find layer was "undertaken with extreme attention to detail at a mineralogical institute in Germany, on the basis of a sample. The find layer is more than 110,000 years old, a time that is generally recognised in North America as the start of Wisconsin's last glaciation. The find layer in Stone Creek is underneath the Wisconsin glaciation" (Hoening, 1981, p. 216f.) and is therefore officially older than 110,000 years.

Prof. R.G. Forbis of the Department of Archaeology at the *University of Calgary* recommended, in his letter of 13 September 1977 (Hoening, 1981, p. 229): "This summer I showed the specimens from Stone Creek to some specialists in lithic technology … He was able to pick out from the sample those artifacts which come from the terrace and felt that they could be identified as products of human manufacture … Probably the same statement could be made of some of the specimens from Choukoutien". These artefacts from China are supposed to be 460,000 years old.

In addition, he confirmed that a piece of bone carbon from the lowest find layer had been subjected to radiocarbon dating in Chicago. The result was that the age was outside the measurable range, so had to be more than 50,000 years old: Forbis "know of no method for obtaining any reasonably precise age for it". Later on he distanced himself from the finds two years later, saying that he "wasn't really interested in the Old Stone Age ".

Following an osteological analysis on 2 September 1980 (Büchner, 1981), the "Lippsche Landesmuseum" in Detmold classified the human bones found at the site as *Homo sapiens*, as they were robuster and larger than those of modern Man (*Homo sapiens sapiens*). Remains of giant beaver (Castoroide) and giant sloth (Paramylodon) were found alongside the human bones. The giant beaver, which grew to lengths of 2.5 metres, lived in North America for 2 million years and supposedly became extinct 10,000 years ago. The fossilised human footprint, which was found directly below the (allegedly) 100,000 year old Wisconsin rubble within the even older sand–lime brick layer, belongs to the same period. It is blackened on the upper side, i.e. the contact side, like all of the find material at Stone Creek – a total of 200 pieces – with a coating of manganese hydroxide, whilst the underside of the stone layer is sand coloured, i.e. light coloured in appearance. The footprint of the primeval American is located within the black area of the stone (Hoening, 1981, p. 260ff.).

Ernst Hoening's evidence of the existence of humans at Stone Creek over 100,000 years ago, also speaks for the presence of a Neanderthal, indeed a Neanderthal forerunner in America. In this case, the history of human origin and settlement in America needs to be thoroughly rethought, yes, rewritten, because the officially non–existent "American Neanderthal" would have lived at the same time as his European brothers.

The fossilised footprint is, however, a far greater sensation which went completely unrecognised by Hoening, because Dr. Martin Büchner's geological evaluation (1981) confirms a geological situation which I documented a long time ago, which is that in America, solid limestone, sandstone, or sand–lime sheets, only a few centimetres thick, whose upper surfaces sometimes appear to be coloured black often rest on granular soil. Moraine–type layers of granular soil are, in turn, found on top of the solid stone, which must come from mudslides, since they never occur in glaciated regions. In the present case, the so–called Wisconsin Moraine stretches from the earth's surface to a depth of 11 metres above the sand-limestone layer containing the footprint.

The geological evaluation related to the stone sheet with the human footprint which lay under the "moraine". But a footprint cannot be made in solid stone, only in a soft surface. The stone can therefore only have hardened after a human walked across the soft sand slurry. Nevertheless: "Hardened deposits dating back to the Ice Age (as in the case of the footprint, HJZ) should not be expected as these layers are generally present as granular soil sediments due to their low geological age" (Büchner, 1981). Even though Martin Büchner draws attention to a paper by E.Th. Sepharim on a pre-moraine rubble conglomerate with glacial grinding (Sepharim, 1973), I see a contradiction, because neither a gravel conglomerate nor a fine gravel layer such as at Stone Creek can be ground so smooth by a glacier that a massive stone sheet is formed out of and on top of granular soil sediments. Absolutely impossible!

I draw attention to the points I make in book *Darwin's Mistake* and explain the described situation with my "Natural Concrete Theory". The thin, hardened and apparently ground layers did not petrify because of pressure from glaciers, but through hydraulic processes. A hydraulic binding agent (lime, grit cement, calcite cement, plaster, or similar) and minerals which are dissolved in the pouring floodwaters, are simply added to the grit, sand or conglomerate. The hardening of the granular soil in question – whether sand, grit or gravel – then only occurs in the upper area and, through a "cementation process", forms a firm but thin rock layer, only a few centimetres thick, depending on how far the binding agent penetrates, on which the ripple markings of the sea floor are often preserved. These ripple markings attest to the formation of thin, hard stone layers under water!

Lasting human footprints in stone layers are produced in these cases in soft original material, as is the case with ash–drift, and not in granular soil. Depending on the type and properties of the binding agent, hardening occurs quickly – in geological terms like lightning – as if superglue had been used. In the case of Stone Creek, we are talking about green sandstone without fossils, whose finely crystalline, carbonatic matrix combined with sharp–edged angular grains of sand, hardened into a limestone sheet inter alia through the addition of potash feldspars and plagioclases.

The footprint must have been left at a time when this mass was still soft, and was then preserved by the rapid hydraulic binding (hardening) of the calcium sandstone. The mudslide known as the "Wisconsin Moraine" can only have pushed itself over the calcium sandstone (with its preserved footprint) when everything had hardened – otherwise it would all have been destroyed! For the same reason, glaciers cannot have hardened the granular soil and then ground it.

The present calcium sandstone with the footprint is geologically classified as green sandstone, because it contains glauconite in the form of blue–green to black potassium iron silicate. But: "glauconite signifies a marine formation environment, so is almost exclusively to be found in marine deposits … It does not belong to the glacial deposits of the mainland …" (Büchner, 1981, p. 263 – expert report, p. 2). This confirms my view that the thin calcium sandstone layer on top of the fine gravel was not hardened by a glacier but was formed under water on a hydraulic basis.

"The black manganese hydroxide coating of the upper surface area is … identifiable as an infiltration of the interstices close to the surface between the grains of sand" (ibid, p. 263 – Evaluation, p. 2), which means they formed when the sediment layer was still soft. This layer hardened before the alleged Ice Age came. The retrieved human relics, which also bore this black colouration, have therefore nothing to do with the "Wisconsin Moraine" in terms of time. The "moraine" must be the result of a mud flood, under which the bones, artefacts and footprints were buried. How old is this "American Neanderthal" really? Simple answer: He is exactly the same age as the calcium sandstone sheet. But how old is that?

Dr. Martin Büchner writes (ibid. p. 263 – Evaluation, p. 2): "The microscopic image of the present sandstone resembles very closely the example of Cretaceous green sandstone of the Ft. Augustus Formation in Alberta, Canada, shown by Pettijohn, Potter and Siever (1972, Figs. 6–30, p. 230) … Such an accumulation of glauconite as in the present calcium sandstone may also be found in the German deposits of the Cretaceous and gives the stone – as in Canada – a green colour".

Now let's read about green sandstone (glauconitic sandstone) in the textbook *Elemente der Geologie* (Credner, 1912, p. 268): "The cement is chalky, marly or clayey. The main development of green sandstone took place in the Cretaceous", i.e. in the age of the dinosaurs. In this connection Dr. Hermann Credner speaks about cement (= binding agent), also in the form of lime. And it is exactly this which is present in the green sandstone at Stone Creek in Canada because, according to Martin Büchner, the stone has a unevenly formed external surface with a white coating (lime-sinter). This layer from the dinosaur period (Cretaceous) hardened with the footprint contained therein relatively fast, like a lime (cement) sand mixture and not extremely slowly over long geological periods and – according to geological dating – this happened during the lifetime of the dinosaurs.

I think that the connections I have outlined are simple to understand. But then it is equally easy to understand that with correct geological dating, dinosaurs and humans coexisted, because the green sandstone layer containing the human footprint, dates back in geological terms to the Cretaceous. This is the most recent of the earth's three ages, which ended 65 million years ago with the extinction of the dinosaurs. But then, giant sloths and giant beavers must have coexisted with dinosaurs as well, because their fossilised bones, just like the finds at Stone Creek and the footprints, are coloured black by manganese hydroxide.

Was the human footprint made in the Cretaceous, as it would appear from the geological dating of the green sandstone, or did this stone harden under water perhaps only a few thousand years ago? But then the whole geological-biological chronology is wrong and the long geological periods are pure illusion. In that case, the ages of dinosaur finds in green sandstone must also be dramatically reduced, at least to the Neanderthal era.

Man Before The Dinosaurs

On 9 June 1891 the publisher of the local newspaper in Morrisonville, Illinois, S.W. Culp was filling her coal bucket. Since one of the lumps of coal was too big, she chopped it up. It broke into two almost identically sized pieces. A delicately wrought gold chain of around 25 centimetres in length, "of old and wondrous workmanship" was revealed (Morrisonville Times, 11 June 1891, p. 1). The ends of the chain which were close together were still firmly embedded in the coal. There, where part of the chain had come free, a round imprint in the coal could be seen. The jewellery was obviously as old as the coal itself. Analysis showed that the chain was of 8–carat gold and weighed 12 grams.

When the owner of the chain died in 1959, the chain disappeared. There are no indications as to the origin of the chain based on its workmanship. The coal in which the chain was embedded is allegedly 260–320 million years old. If we accept that this often described case is authentic, the consequences are unbelievable: was there a culture in this primeval, pre–dinosaur epoch, which could produce gold chains? Then the theory of human evolution would be the biggest mistake of the second millennium.

Fig. 22: Clay Figure. Retrieved from a depth of 100 metres.

The other solution – again and as always – is to put the wrong date on when the coal was formed. Did coal, generally speaking, form only a few thousand years ago rather than hundreds of millions of years ago during the Carboniferous? In this case, the presence of a gold chain in a lump of coal presents no problem. However, the 300 million years of the geological timescale then appear to be a purely imaginary phantom period.

The production of a gold chain is a job for an expert and certainly not the work of a "Stone Age man". The oldest known gold chains are around 5,000 years old. 8-carat gold is an alloy which is produced using eight parts gold and sixteen parts other metals, usually copper. But there has never been an 8-carat standard. When the Morrisonville chain was discovered gold alloys were usually 15-carat and bore a stamp.

This find was no isolated case. Other finds in coal from the Carboniferous include:

• A type of measuring beaker in Wilburton, Oklahoma in 1912. While working the coal, Frank J. Kenard broke a large piece apart and a type of pot or measuring beaker made of iron fell out. This find was witnessed by Jim Stull, an employee of Municipal Electric and a notarised declaration was made before Julia L. Eldred.

• A thimble (J.Q. Adams in American Antiquarian, 1883, pp. 331–332).

• A spoon (Harry Wiant in *Creation Research Society Quarterly*, Issue No. 1, Year 13, 1976)

• An iron cauldron and human footprints in coal (Wilbert H. Rusch in Creation Research Society Quarterly, Year 7, 1971).

• An iron instrument (John Buchanan in Proceedings of the Society of Antiquarians of Scotland, Year 1, 1853).

There are also finds from even older geological layers:

• In 1844, Sir David Brewster presented a report to the British Society for the Advancement of Science. He sid that workers at a stone quarry in Kingoodie near
Dundee, Scotland, had smashed up a block of sandstone. The head of a nail, three centimetres of whose shaft was still embedded in the rock, was seen (Brewster, 1845).
The sandstone in the area in question is supposedly at least 387 million years old, having formed in the even older (lower) Devonian, the period before the Cretaceous.

• According to a report in the magazine *Scientific American* on 5 June 1852 (p. 298) a metal ship or vessel with silver inlay was found in geological layers which were far too old.

• A gold thread was found embedded in pure rock near Rutherford Mills, England (*The Times, London*, 22 June 1844, p. 8 and *Kelso Chronicle*, 31 May 1844, p. 5).

• In 1851 in California a broken iron nail was found in a lump of quartz. The London Times reported the find (24 Dec 1851, p. 5) under the headline "A Puzzle for the Geologists".

• Rene Noorbergen (1977) reported on the find of a metal screw in the US State of Virginia. It was enclosed in a round, mineral hollow form (a geode).

• In 1889 in09 Nampa, Idaho, a small artistically shaped clay figure was found, representing a person wearing clothing (Fig. 22). This artefact was discovered at a depth of 100 metres when a well was being drilled. Prof. F.W. Putnam pointed out that the surface of the figure had a ferrous encrustation and that a red coating of iron oxide is still partially preserved (Wright, 1897, pp. 379–391).

• "In the 16th Century the Spanish found an 18cm long iron nail inside a rock in a Peruvian mine; it was without doubt many thousands of years old. In a country where iron was almost unknown, this discovery was rightly seen as sensational. Francisco de Toledo the Viceroy of Peru accorded the nail a place of honour in his study" (Thomas, 1969).

• "Platinum ornaments were found on the coast of Ecuador in South America. This tiny piece of news throws up a big problem for science: how could the inhabitants of pre–Columbian America have produced temperatures of around 1,770°C, when the European only managed to do this two hundred years ago?" (Mason, 1957).

Footprints Which Are Too Old

It appears that in many instances people living several hundred million years ago in the earth's middle period, during the age of the dinosaurs, left footprints behind:

• In Kentucky (*Science* News Letter, 10 Dec 1938, p. 372).

• In Missouri (Henry Schoolcraft und Thomas Benton in: *The American Journal of Science and Arts*, 1822, p. 223–231).

• In Pennsylvania (*Science News Letter*, 29 Oct 1938, pp. 278– 279).

• In Nevada. In Fisher Canyon, Pershing County, a shoeprint was discovered. The sole is so clearly defined that even traces of a type of twine can be seen. Samuel Hubbard, Director of the Museum of Archaeology in Oakland, California, believes that in 1927 it was not yet possible to produce this kind of shoe. On the basis of geological dating of the coal layer, the age of this print was estimated at 160–195 million years.

• In 1983 the *Moscow News* (No. 24, p. 10) reported on the find of what appeared to be a human footprint in 150 million year old stone in Turkmenistan, right beside the enormous fossilised three–toed footprint of a dinosaur. Prof. Amannijazow, the corresponding member of the *Turkmenian Academy of Sciences* conceded that the print resembled that of a human foot, but that he did not regard this as proof of the coexistence of humans and dinosaurs.

• The members of a Sino-Soviet paleontological expedition in 1959 found "in the Gobi Desert in a stone buried beneath the sand, the print of a million year old shoe, which came from a time when there were as yet no humans" (*Moscow magazine Smena*, No. 8, 1961).

• In the specialist journal *American Anthropologist* (Vol. IX/1896, p. 66) the finding of a perfect footprint, around 37 centimetres in length, four miles north of Parkersburg, West Virginia, is described. According to modern geological dating, a human must have been running around the eastern USA 150 million years ago at the time of the dinosaurs.

• Several foot and shoe tracks were discovered during the 1970s in Carrizo Valley in northwest Oklahoma. The 52 centimetre (!) long tracks were not only in the Morrison Formation which is typical for dinosaur finds but were also directly beside dinosaur footprints in the same stone layer. Other tracks were found in Cretaceous Dakota sandstone.

Human footprints from the period before the dinosaurs were also found in further US states: the head of the Department of Geology at Berea College in Berea, Kentucky, Prof. W.G. Burroughs (1938) wrote in *The Berea Alumnus* (November 1938, p. 46f.) of "creatures who at the start of the upper Cretaceous walked on their two hind legs, with feet which resembled those of humans and who have left their traces on a beach in Rockcastle County, Kentucky. It was the time of the amphibians, in which animals moved on four feet…and whose feet were in no way similar to those of humans; but in Rockcastle, Jackson and several other places between Pennsylvania and Missouri creatures with feet existed, whose feet strangely resemble those of humans and who walked on two legs". The author of these lines proved the existence of these creatures in Kentucky. With the assistance of Dr. Charles W. Gilmore, the Curator of the Department of Palaeontology in Mammals at the Smithsonian Institute, it could be shown that similar creatures also lived in Pennsylvania and Missouri.

It was further established by the scientists that the grains of sand within the tracks were closer to each other than they were outside, owing to the pressure which is transmitted by the weight of the body via the feet into the ground. The grains were closest in the heel area as the pressure here is greater than at the front of the foot.

The contention that Indians could have carved the tracks into the stone (*Science News Letter*, 1938, p. 372), was rejected by the sculptor, Kent Preiette: There were no indications of chiselling or cutting work of any kind to be found either in the enlarged micrographic photographs or in the enlarged infrared photographs (*Courier-Journal Magazine*, Louisville, Kentucky, 24 Mai 1953).

Burrough's conclusion was that humanoid footprints were pressed into wet, soft sand before this solidified into hard stone around 250 million years ago. As such, human creatures must already have lived during the earth's middle period during the age of the dinosaurs.

An official reaction followed in *Science News Letter* (1938): Human–like footprints in stone puzzle the scientists. They cannot be human because they are far too old – but what strange two–footed amphibian could have made them? We may wish the orthodox experts much fun in this search which has been fruitless since 1938 and will remain that way.

The following statement exemplifies geology: "What can't be, shan't be". This attitude was categorically reinforced in *Scientific American* (Vol. 162, 1940, p. 14):

If Man or only his apish ancestor or even early mammal forerunners of this ape–ancestor in whatsoever shape had existed in such a distant time as the Carboniferous then the entire science of geology would be so fundamentally wrong, that geologists should pack in their careers and go and drive trucks.

Perhaps orthodox geologists really should apply for their driving licences just in case, because there are also human bone finds in layers which are far too old. In *Heimatliche Plaudereien aus Neunkirchen* (Local Talks from Neunkirchen) which was sent to me by Manfred R. Hornig, a report from 1975 (p. 40) reads: "1908 Visit of the international study commission for the examination of prehistoric finds of a fossilised human lower leg bone in the eastern Braun seam, 2nd level, cross–cut 3. 1909 Transportation of the 'Braun' find to the Prussian State Museum in Berlin (secret)". This matter had to be treated as secret because humans could not have lived in the Carboniferous maybe 300 million years ago.

In December 1862 the journal *The Geologist* published an interesting report: In Macoupin County, Illinois, the bones of a man were recently found on a coal–bed capped with two feet of slate rock, 90 feet (27.5 metres) below the surface of the earth … The bones were covered with a crust or coating of hard glossy matter, as black as coal itself, but when scraped away left the bones white and natural.

The coal mined in the middle of the 19th Century in Macoupin County is, however, 320–286 million years old (Cremo/Thompson, 1997, p. 346). According to geological datings, this man must have lived before the dinosaurs.

In my book *Mistake Earth Science* an unusual fund is documented which existed and can be examined. At his house in Bogotá, Colombia, Prof. Jaime Gutierrez Lega showed me a photograph of a fossilised hand, which was discovered in an area of dinosaur relics (Zillmer, 2007, picture 73, p. 237). On the occasion of the "Unsolved Mysteries" exhibition in Vienna, Austria, I had this find – which was introduced by me to the world for the first time – flown in from Colombia, exhibited and examined: on the basis of the embedded fossils the stone can indisputably be dated geologically to the age of the dinosaurs.

Hofrat (Austrian honorific) Dr. Reinhart Fous (Chief Physician at the Federal Police

Headquarters in Vienna) and Prof. Dr. Friedrich Windisch from the Institute of Anatomy of the University of Vienna came to the conclusion that this was the right foot of a hominid and a hominid hand. These experts based their findings on individual bones which are only present in human extremities. Conclusion: parts of skeletons which are indisputably human have been found in stone which unarguably dates back to the earth's middle period. So: dinosaurs and humans or their forerunners (hominids) co-existed – a scientifically proven fact which anyone can check.

Or, to save conventional science's face should we be looking not only for amphibians which walked on two legs and had humanoid feet but also for dinosaurs which possessed a specifically human bone.

Then again, perhaps the scientists are searching in vain, because the scientific front is crumbling: The first primates probably appeared about 80 million years ago and looked dinosaurs in the eye (*Nature*, Vol. 416, 18 Apr 2002, pp. 726–729). And on the basis of having studied fossil DNA, the American biologists Sudhir Kumar and Blair Hedges believe that most mammal species already existed 100 million years ago at the time of the dinosaurs (*Nature*, Vol. 392, 30 Apr 1998, pp. 917–920).

In the phosphate rocks of South Carolina, an enormous mass grave containing land–living mammals (incl. mammoths, elephants, pigs, dogs and sheep) lying side by side with birds and aquatic animals (incl. whales and sharks) was discovered. Human relics were also unearthed (Willis, 1881). Prof. F.S. Holmes, palaeontologist and curator of the National Historic Museum in Charleston, documented the find of a six metre–long lizard in a report to the Academy of Natural Sciences. He noted that this find dated to the late Tertiary, "when the American Elephant, or Mammoth, Mastodon, Rhinoceros, giant ground sloth (Megatherium), Hadrosaurus and other gigantic quadrupeds roamed the Carolina forests" (Holmes, 1870, p. 31). In other words, an expert described the finding of Hadrosaurs which allegedly lived 80 million years ago (by today's reckoning) together with mammals which are more than 50 million years younger, and which were even in a mass grave with humans. On the title page of his book The Phosphate Rocks of South Carolina, a Hadrosaurus skeleton is clearly depicted. Plesiosaurs were apparently also discovered. It has been confirmed that dinosaur skeletons from the Upper Cretaceous were found together with large mammals, at least 30 million years younger and even younger humans in a single enormous mass grave which had been washed together. It seems that large mammals, humans and dinosaurs died at the same time during a major natural catastrophe. These phosphate deposits have now disappeared, having been exploited to exhaustion. If large mammals and dinosaurs did coexist then one should also find footprints in these layers. In Uzbekistan, 86 consecutive hoof tracks from horses were discovered in a layer which geologically dates back to the time of the dinosaurs (Kruzhilin/Ovcgarov, 1984). In a paper by the U.S. Geological Survey, 1,982 photographs of the Grand Canyon were published which showed horse–like hoof tracks. The problem is that this layer has been geologically dated to 100 million years before the first ungulates appeared in evolutionary history (Geological Survey Professional Paper 1173, Washington D.C., 1982, pp. 93–96, 100). Similar hoof tracks have been found alongside a thousand dinosaur tracks in Virginia (*Science* News, Vol. 136, 8 July 1989, p. 21). The coexistence of more highly developed mammals such as ungulates and dinosaurs however, completely contradicts the principles of geology and evolution! Conclusion: large mammals, humans and dinosaurs coexisted. This coexistence which is proven by hard facts contradicts the propagated ladder of (macro)evolution.

Tertiary Man

Between 1912 and 1914, the respected Argentinean palaeontologist, Florentino Ameghino discovered stone tools, hearths, broken mammal bones and a human vertebra in a 5–1.7 million year old Pliocene layer at Monte Hermoso, Argentina (cf. Cremo/Thompson, 1997).

In order to ensure a correct dating of the implements, Florentino Ameghino invited four renowned geologists to examine them. The team of experts confirmed: "All of those present stated that the stone artefacts…were in intact, undisturbed terrain and found in their original positions … They were found in situ and should therefore be regarded as having been fashioned by humans, contemporaneous to the geological level in which they were found…these people lived in a time which coincided with the "Chapadmalal Phase" (Roth et al., 1915, p. 422f.). Dating of this Chapadmalal Formation has put its age at 3–2.5 million years (Anderson, 1984, p. 41) or 3–2 million years (Marshall et al., 1982, p. 1352).

Ameghino's brother, Carlos (1915, p. 438f.) discovered a range of stone tools, traces of fire and the upper thigh bone of a Pliocene Toxodon (cf. Zarate/Fasana, 1989) all in the same geological layer at Miramar, Argentina. Toxodon, an extinct South American ungulate resembled a hairy, short–legged hornless rhinoceros.

In the Toxodon's upper thigh bone, Ameghino found a stone arrowhead. The almost complete Toxodon back leg, with intact articulation, was clear proof that this leg had not been moved since it was embedded in the geological layer. At the time of the discovery, it was not yet known that this animal had only become extinct in South America a few thousand years before.

However, Carlos Ameghino already differentiated between the younger and large Toxodon burmeisteri and the Toxodon chapalmalensis from Miramar: These "Toxodon bones are of a dirty–white colour, as is typical for this geological layer and not blackish as might be expected if they had been in contact with the magnesium oxides of the (younger) Ensenadas" (Ameghino, 1915, p. 442). The bones were also full of Chapadmalal loess.

This find confirmed that culturally advanced humans lived in Argentina at a time when, according to conventional teachings, human beings had only just started to evolve in Africa, in the shape of Australopithecus (Lucy).

At the beginning of the 20th Century, an influential group of scientists did all in its power to ensure that supposed proofs of the existence of Tertiary Man were buried once and for all (Hrdlicka, 1912). An international role was also played by the German prehistorian, Hugo Obermaier (1877–1946), whose memorial is the Hugo Obermaier Society at the University of Erlangen–Nuremberg. Antonio Romero (1918) quoted Obermaier's 1916 book Fossil Man in Spain and rejected conclusions about the existence of Myocene and Pliocene Tertiary Man in South America: supposedly the finds were the legacies of modern Indians.

After the discovery of the Chapadmalalian Toxodon upper thigh bone in Miramar, a fully preserved part of a Toxodon spine was found, in which two stone arrowheads were stuck. Boule wrote: These discoveries were disputed. Reliable geologists affirmed that the objects came from the upper beds, which formed the site of a paradero, or ancient Indian settlement, and that they were found today in the Tertiary bed (older) only as a consequence of disturbances and resortings which that bed had suffered (Boules/Vallois, 1957, p. 492).

Those who make such assertions in scientific papers should as a matter of course state their references as well. With regard to the "reliable geologists" however, Boule merely gives a footnote in which he only mentions Romero's 1918 essay in which the (anomalously) old tools were ascribed to modern Indians. The geologists' report, however, goes unmentioned. Worse yet, Boule credits Romero's 1918 esssay without verifying it in any way, although the latter's geological views had already been shown to be false following the Bailey Willis examinations (Hrdlicka, 1912, p. 22f.).

In his statements about the Miramar finds, Boule provides a classic case of prejudice and preconception masquerading as scientific objectivity. For example, Boule said nothing at all about the abovementioned discovery of a human jaw in the Chapadmalal Formation in Miramar (Cremo/Thompson, 1997, p. 289).

E. Boman, whom the critic Boule regular brings on as an authority, also visited the site. He several times mentions the possibility of fraud, which he didn't rule out, but found that: In the final analysis there undoubtedly exists no conclusive proof of fraud. On the contrary, many of the circumstances speak strongly in favor of their authenticity (Boman, 1921, p. 348).

But now let's take another look at criticism with regard to the finding of human relics from the Tertiary (65–1.7 million years). Antonio Romero visited the area of Miramar and was shown the fairly new stone tools from the paraderos (settlements) of the coastal Indians. These showed similarities with the objects found in the Pliocene layer in Miramar. Romero was convinced that these had been fashioned by the same makers, as had made those items which supposedly belonged to a far too fantastic epoch (Romero, 1918, p. 12).

Carlos Ameghino himself confirmed that at least since the Chapadmalal…people of the *Homo sapiens* type have existed in this area who, surprising as this may appear, had achieved a level of culture comparable to that of the newer prehistoric inhabitants of the region (Ameghino, 1915, p. 449).

The stone weapon tip found in the upper thigh bone of the Toxodon has been sharpened along its length on one side and retouched to a point at both ends and it is roughly in the shape of a laurel leaf. This Argentinian leaf–shaped tip very closely resembles those Solutrean leaf points, which became known as "laurel leaves" (Ameghino, 1915, p. 445). The newer cultural stage of the Old Stone Age (Solutrean) took its name from the find site of Solutré in Burgundy, France, and is famous for its retouched blade and notched weapon tips. This period is officially said to have lasted around 5,000 years, ending 17,000 years ago.

Romero's criticism (Romero, 1918) is, in principle, justified, because at the time when the ape–like creature, Lucy, had just started to toddle in Africa and humans had, so to speak, just started to evolve, how could finely-wrought stone projectiles of high technical quality be produced in South America? And then no further development in tool and weapon manufacture for 3–2 million years, while the evolution of ape to man took place in Africa?

Nonetheless, this Toxodon which was hunted by Man in Argentina should be regarded as an authentic case and not a fake, particularly since, at the beginning of the 20h Century, the finds fitted into the geologists' world view which was then based on old scientific principles – unthinkable today. How does it all fit together?

Quite simply: the dating is wrong! Between the layers of the Upper Pliocene in South America and the layers with the Solutrean weapon tips in France, there is a gap of 2 million years, which has to be closed. If we reduce the Miramar layer which dates to the newer Tertiary by this length of time, there is no longer any contradiction. Contrary to the official geological timescale, earth layers do not represent any sort of clock. These layers are formed by rapidly occurring natural cataclysms and flooding or by major meteorite impact, which hurls the earth material from one place to another. By transporting material, landslides, volcanic eruptions or tsunami waves which penetrate deep inland also create new layers in different places. The material of these layers is not new, but the layer is newly formed and contains animal remains or human relics which have been washed into the material.

Anyone who can follow my argument must also question the allegedly great age of these sediment layers, whether of granular soil or of rock. A reduction of geological times by whatsoever factor means, however, that the sacrosanct geological timescale and its inseparable biological counterpart, the evolutionary timescale, cannot be correct.

Prof. Dr. Bazon Brock rightly asks: "The basis for all models of creation processes (our solar system, our planet, life) is time, which we use for our models of evolution. The fairy-tale formulation 'Once upon a time …' in itself illustrates how we use measurements of time which are beyond verification and the bounds of imagination, to get rid of all the difficulties we have with our conceptual models in the vague, unimaginable depths of time. That is the real fairy story" (Brock, 2001, p. 14).

Now let's look at some human finds from the Tertiary in Europe which are also anomalously old. In the Pliocene (–5 to 1.7 million years) towards the end of the Tertiary, warm sea waves broke on the southern slopes of the Italian Alps and left many coral and mollusc deposits.

In the late summer of 1860, the Italian geologist, Prof. Guiseppe Ragazzoni found fossilised mussels around 10 km southeast of Brescia near Castenedolo, which were located in open Pliocene layers at the foot of the Colle del Vento mountain. To his surprise, Ragazzoni found fossilised human bones in the old marine deposits. First he was suddenly holding a piece of a cranium in his hands "which was completely full of fossilised corals and covered with the blue–green lime (kink) typical of this formation". Astonished, Ragazzoni looked further and found other bones from the ribcage and limbs "which obviously belonged to a human being" (Ragazzoni, 1880, p. 120).

Since geologists who were brought in to give advice believed that human bones could not possibly be present in such an old layer and concluded that they came from a very deeply-dug grave, Ragazzoni threw the bones away "not without regret". But the story doesn't end there.

On 2 January 1880, around 15 metres away between the coral bank and the seashell–clay layer above it, further human bones were discovered. Prof. Ragazzoni was informed and together with an assistant, excavated the bones himself. He found a large number of skull, vertebra and rib fragments, teeth, arm and thigh bones, shins and fibulae as well as a tarsal bone and two toe bones. On 25 January 1880, at a distance of 2 metres at the same depth, numerous other bone fragments were retrieved, including a large number of cranial fragments.

All were completely covered in and full of clay, small seashells and coral fragments, which removed any suspicion that the bones had come from a burial site; on the contrary, this confirmed that they had been washed up by the waves (Ragazzoni, 1880, p. 122).

Three weeks later, a complete skeleton was found in the middle of the layer of blue clay. The stratum of blue clay, which is over 1 metre thick, has preserved its uniform stratification and does not show any sign of disturbance (ibid. p123). The slowly deposited layers were carefully removed and the entire skeleton was freed (ibid. p. 122). Unlike the finds in 1860 and those made this year (1880), this intact skeleton appeared in the middle of the kink layer…over which a layer of yellow sand had settled. The layered clay removed all doubt that the skeleton had been washed up in more recent times. The kink was in a condition "which ruled out any rearrangement by humans" (ibid. p. 123).

The cranium was restored by the anatomist, Prof. Guiseppe Sergi of the *University of Rome*. He could not find any difference between it and the skull of a modern woman. In his report he wrote: "They (the Castenedolo skeletons) are an indisputable document of the existence of Tertiary Man – not a forerunner, but a person, of completely human character" (Sergi, 1884, p. 315).

However, modern humans are supposed to have appeared in Europe 40,000 years ago at the very earliest. The kink layer in which the Castenedolo skeletons were found has been dated by several geologists as belonging to the Astian period (Oakley, 1980, p. 46), i.e. the middle Pliocene. As such, the existence of modern Man in Europe can be dated to a time 4–3 million years ago, when Lucy was still toddling around in East Africa. Such finds reveal human evolutionary history to be a myth and the geological age datings of the layers must also be questioned. But in this case, everything becomes younger, because all the layers are wrongly dated. Following the comment that the Castenedolo bones were anatomically modern, Prof. R.A.S. Macalister wrote in 1921 (p. 184f.): "If the Castenedolo bones really belonged to the stratum in which they were found, this would imply an extraordinarily long standstill for evolution. It is much more likely that there is something amiss with the observations. The acceptance of a Pliocene date for the Castenedolo skeletons would create so many insoluble problems that we could hardly hesitate in choosing between the alternatives of adopting or rejecting their authenticity".

It was not only this find which was denied. Even as recently as 1969, experts tried to discredit these finds and claimed that their age was lower. Eighty nine years after their excavation, they used radiometric and chemical tests. The test methods were shown to be faulty as contamination of the bones by dirt, air and micro–organisms during their 89–year museum storage, could not be ruled out. The effects of acid and decay during the long period of embedment in the marine sediments are further unknown factors, which falsify the readings.

A radiocarbon test of the Castenedolo bones showed them to be only 968 years old. However, the method used for the dating is now considered unreliable. Furthermore, the bones contained a level of fluorine which was far too high for extant bones (Oakley, 1908, p. 42). The unexpectedly high concentration of uranium also suggested a great age.

As such the Castenedolo case demonstrates very effectively the inadequacies of the methodology applied by the paleoanthropologists (Cremo/Thompson, 1997, p. 340). You can't have your cake and eat it: either the existence of modern humans during the Tertiary is a fact, or an attempt is being made to classify these fossilised bones as relics from the last post–Ice Age millennia, the Holocene.

The excavation reports, however, show clearly that the bones lay in undisturbed Pliocene marine sediments. If the Castenedolo bones are from the Holocene as suggested by radiocarbon dating, then the corresponding layers of the Middle Pliocene must also be made 3–4 million years younger. Either one or the other! The second solution would correspond with my interpretation because the marine sediments did not form in the Pliocene, but only a few thousand years ago through natural catastrophes during the Deluge – the geological dating is quite simply wrong.

A Change of Opinion

In 1913, Prof. Hans Reck of the University of Berlin discovered a human skull in the Olduvai Gorge in East Africa. The skeletal remains, including an intact skull, were firmly cemented into the matrix and had to be released using hammer and chisel. In view of the fossils lying immediately beneath the find, the Reck skeleton (from Bed II) was dated to the older Pliocene (1.7–0.72 million years). Louis Leakey supported this dating (*Nature*, 1931, Vol. 121, pp. 499–500). This opinion was reinforced in 1931 by the discovery of stone tools in the Olduvai Beds I and II. Today, it is conceded that Bed II is 1.15 million years old.

In a letter published in *Nature*, Leakey, Reck and A.T. Hopwood (of the British Museum of Natural History) said that as stated by Reck, the skeleton had been lying in Bed II from the outset. The layer sequence as described at the time is still acknowledged today: the first four layers comprise various types of volcanic tuff deposited in the water, whilst Bed V above is a type of loess.

After heated debates in *Nature*, Reck and Leakey finally withdrew their assertion and said that the skeleton had probably entered Bed II at some later date and was not older than the continuity break below Bed V. The reason they changed their opinion is unknown. Were their academic reputations at risk?

The reason for this dispute was that the man in the Olduvai Gorge was not a Neanderthal, but belonged rather to the Aurignacian type (MacCurdy, 1924, p. 423). This means that it was an early–modern human, whose skeleton does not differ anatomically from our own. But modern Man has existed officially in the Near East only for 100,000 years and in Africa for 140,000 – or, more recently, 200,000 years. If the evolutionary ladder of human ancestry is to be seen as correct, then there naturally cannot have been a modern human (*Homo sapiens sapiens*) over 1–2 million years ago. And that's that! So for reasons of dogma, the find must have been buried in a layer which was far too old, although the experts had to chisel out the embedded skeleton from the undisturbed matrix. If *Homo erectus* had been found in Bed II, dating his age at a million years would not have raised even a whisper of protest.

In 1960, a new surface find was made in the Olduvai Gorge: the skull was classified as *Homo erectus* (OH 9). In fact, this species of human should not be found in newer, surface layers but in those which are deeper and older. In other words, exactly this skull would have fitted in perfectly to the time determinations laid down for Bed II and thus also with the human evolutionary ladder. And that's what happened! Because allegedly remains of the Bed II matrix were stuck to the base of the skull, this surface find was classified as coming from the far deeper Bed II – and dated to a million years old. This layer exactly fits in to where *Homo erectus* is supposed to be in terms of time. Overall, this is a perfect example of how a round peg is made to fit into a square hole.

Anthropology has solved the contradictions with the latest scientific expertise and genial simplicity for the experts: for an orthodox expert, putting skeletons into the correct geological layers is a simple exercise. The uninitiated people and laymen just have to believe him. If they don't they are incorrigible ignoramuses. The world–renowned anthropologist and star professor of the Johann–Wolfgang–Göthe University in Frankfurt, Germany, Reiner Protsch, acted according to these principles and confirmed that it was modern humans who had been found in the anomalously old Bed II layer: "Theoretically, several facts militate against these hominids being very old, for example, the morphology" (Protsch, 1974, p. 382). Protsch also went by the motto: a modern human must be young.

But Prof. Protsch was ostensibly able to bolster his opinion scientifically. After 61 years, the Reck bones were, allegedly, brought out of the dusty museum cellars and dated using the radiocarbon method. The skeleton of the modern human from the geologically dated million year old Bed II, was said to be around 17,000 years old (Protsch, 1974) and this dating fitted exactly into the time horizon for modern humans. Thus the retrospective proof was provided scientifically, using modern measuring techniques, to show that a mistake had been made by the experts of the time and that a reburial in old layers had taken place.

Protsch's dating was already criticised at the time because the accompanying circumstances of the dating appeared unreliable. In addition, various radiocarbon datings of finds from Olduvai had already been deemed far too young by the experts. These figures, which were too low to fit in correctly with human history, were explained through contamination by secondary carbon compounds from within the earth. Since the expected result was already known prior to the datings, incorrect readings simply had to be ignored or just correctly interpreted, i.e. defined as being outside the measurable range or mathematically adjusted. The imagination of experts knows no bounds: the end justifies the means because human evolution is already proven. So what difference do a few mistakes make?

In keeping with this pattern, Protsch performed a service to the experts for which they had been waiting longingly and desperately: he shed light on darkness and solved a very problematic discovery using modern research. Suddenly, the Reck skeleton fitted onto the evolutionary ladder as if tailor–made. The case was finally solved and closed for the experts. The saviour was celebrated all over the world. Further discussion was prohibited. But what became of Prof. Dr. Dr. Reiner Protsch? This is the same Reiner Protsch mentioned in connection with the falsification of age determinations for Stone Age skulls who, since 1991, bears the aristocratic title of "von Zieten".

Was Protsch a dating expert? According to a report in *Der Spiegel* (34/2004) he was barely able to use the radiocarbon dating equipment (C–14 instrument). It was the physicist Bernhard Weninger, who transferred to Frankfurt in 1981 and got the measuring site ready: "It looked great but the laboratory had no calibration parameters, the properties of the counter were completely unknown, it had never been used before I arrived". It seems that in 1974 super–Professor Protsch was not able to carry out a professional radiocarbon dating and quite simply invented the younger age for the Reck skeletons from the Olduvai Gorge, so longed for by the experts.

"Internally, the C–14 professor was soon infamous for cheating like this. His assistants spoke of 'protsching' and 'mental dating'. In this way important fossils were placed in completely wrong millennia. The allegedly 36,000 year old 'Neanderthal of Hahnöfersand', for example, really died 5,500 years before Christ" (*Der Spiegel*, 34/2004).

But Protsch is not a case of a single scientist, hungry for fame and money, i.e. a loner. He is part and parcel of the cheating accepted by the congregation of paleoanthropologists with regard to human evolution and, ultimately, geological dating as well.

Similarly, the German society "AG Evolutionsbiologie" – founded of the evolutionary biologists in 2002 in Germany – also served to argumentatively conceal these falsifications in evolution research, with the aim of "limiting the influence of anti–Darwinism in schools and amongst the public" (*press release*, Dr. Georg Kääb, 29 Apr 2004). Accordingly, the combined strength of the united keepers of the Grail of evolution is exerting pressure on the main media, in an effort to prevent the publication of any material which deals critically with evolution.

The chairman of AG, Prof. Ulrich Kutschera was allowed to present his opinions unopposed on the German *ARD* television programme "W wie Wissen" on 20 October 2004, during which he warned against dangerous and popular books such as *Darwin's Mistake*. The result is that publications which are critical of evolution theory are no longer reported and are marked with a lock–flag. Just as the Catholic Church tried, at the end of the Middle Ages, to use experts trained in rhetoric to conceal the Bible's many inconsistencies and simultaneously used all means to silence its opponents – so the aim now is to standardise public opinion. Today, analogous to the situation then, every available means is used to counter the critics of evolution theory. Let us therefore examine some further proofs against evolution – as long as this is still possible in Germany.

Ploughed Up

During the 1849 Gold Rush in California, gold was found on the slopes of the Sierra Nevada in the gravel of old river beds. Soon mining companies were running shafts into the mountain and the gold–bearing gravel was being washed. During this work, stone artefacts and, occasionally, human fossils were found (Cremo/Thompson, 1997).

A table mountain in California's Tuolomne Country on the western edge of Yosemite National Park achieved fame because of its artefacts. The summit of the table mountain comprises a massive, allegedly nine million year old lava dome. Below this and other rock layers, there are river gravel layers containing gold, which are located above the 55 million year old bedrock and are said to be between 55 and 9 million years old. To extract this river gravel, a network of horizontal galleries, up to several hundred metres in length, were driven into the bedrock. From these, vertical shafts then branch off into the lower lying gravel layers. Other mines were run diagonally from the mountain slope into the upper layers of this deposit.

In the compact–hard gravel layer, miners discovered spearheads, ladles with handles and an unusual notched object made of slate, which appeared to be a bow grip. Hardly anything is known about the discoverers, how the finds were made or their stratigraphic positions. A jawbone was also discovered below the basalt sheet of the Tuolomne table mountain (Becker, 1981, p. 193).

An obviously man-made object which appears to be a type of grindstone or a milling implement (Whitney, 1880, p. 264) was retrieved from a wagonload of material taken from inside the table mountain. In 1853, Oliver W. Stevens recovered a mastodon tooth, together with a large stone bead which had been bored through, from another wagonload of auriferous gravel (Whitney, 1880, p. 264). As the finds are from gravel layers which are allegedly 55–33 million years old, one must assume that these artefacts are of a similar age.

In an auriferous gravel layer 54 metres below the surface, Albert G. Walton, one of the owners of the Valentine Mine, found a stone mortar with a diameter of 36 cm (Whitney, 1880, p. 265). A fragment of fossilised human skull was also discovered in this mine. In 1862 a further stone mortar with a diameter of 79 cm was dug out of a gravel layer 60 metres below the surface beneath an 18 metre–thick basalt layer, about 550 metres from the tunnel entrance (Whitney, 1880, p. 266).

Overall, countless artefacts within an area of 160 km were discovered in dozens of mine galleries. That the miners attempted to deceive and falsify over a period of years is therefore out of the question. Or did somebody in the 19th Century already come up with the idea of confusing today's paleoanthropologists as a precaution? Why? At the beginning of the Gold Rush, Darwin's book on the origin of the species had not yet appeared.

As evolution theory was only gradually accepted by science around the turn of the following century, it was only at this time, 50 years after the first finds, which they started to be contested, because Stone Age artefacts which are over 30 million years old are definitely outside the bounds of the imagination for geologists and anthropologists.

Fault was found with the fact that the various stone mortars discovered, did not bear the marks of aging or wear and tear, which must have been there through being transported by Tertiary mountain torrents (Holmes, 1899, p. 471). Since the simple mortars were made of hard andesite, pronounced signs of aging are hardly to be expected. It was also suggested that the stone mortars had been brought into the mine galleries by Indians living in the vicinity (Holmes, 1899, p. 499f.).

This opinion appears unrealistic because from 1849 during the Gold Rush era, the Indians were driven out of the mining region. They can therefore hardly have brought mortars into the mine galleries while the miners were extracting gold there. Nevertheless, in 1908, William J. Sinclair added his objection:

There were clear indications that there had once been an Indian camp in the neighbourhood. After searching for only half an hour a few metres north of the mining company's buildings, a pestle and a flat grindstone were found. Holmes reports similar finds ... (Sinclair, 1908, p. 120). The geologist Prof. J.D. Whitney of the University of California, who acquired possession of many of the Tuolomne County artefacts, had already noted in 1880 that portable mortars, such as were found in the mine galleries, were not used by the Indians living in California at the time of the Gold Rush (Whitney, 1880, p. 279).

Holmes made the point that: Perhaps if Professor Whitney had fully appreciated the story of human evolution as it is understood today, he would have hesitated to announce the conclusions formulated, notwithstanding the imposing array of testimony with which he was confronted (Holmes, 1899, p. 424). In other words: findings which contradict evolution may not go beyond the elitist circle of experts – to avoid arousing uncertainty amongst the public at large.

Let us examine more closely the consequences of artefacts which appear to old. Naturally, officialdom attempts to discredit modern finds in old layers, in order to preserve the theory of human evolution. The argumentation is that the Stone Age artefacts from deep inside various mountains of the Sierra Nevada are identical to those which one can find outside these mountains on the surface.

However, if one examines the tools, both the supposedly old ones and those which are definitely new, one sees that these are simple artefacts, such as were produced all over the world and at all times by cultures of the Neolithic (Young Stone Age) type. This statement must be emphasised. But I wouldn't underline the resulting conclusion: However, if completely different peoples produced similar tools completely independently of each other, it is possible that this may also be the case for people separated from each other ... by millions of years (Cremo/Thompson, 1997, p. 321).

It might be possible, but it isn't likely that similar Neolithic cultures existed over 30 million years ago and then appeared again just a few thousand years ago. Where's the development there? These identical artefacts of very old and very young cultures are lying virtually beside each other after 55–30 million years – inside and outside the mountain. Is it not logical to suppose that these identical artefacts, lying around in the open on the one hand, and enclosed deep inside the mountains on the other, belong to a single younger time phase and settlement phase of the American West? Simply for reasons of continuity, these cultures must have lived relatively recently, since there is no culture which can be verifiably traced back over 30 million years!

Given these considerations, the results of the geological age determinations of Tertiary layers in the table mountains and the surrounding area must be rejected as far too old. The table mountains and the river valleys did not form 30–55 million years ago and the basalt dome from the flowing lava did not form 9 million years ago. These are, rather, water and mud floods which were caused by intense volcanic activity, whereby series of lava flows rolled over the devastated landscape and created the protective basalt dome of the table mountains. This was a short-term catastrophe, standing in for longer geological periods; thus once again, a time-impact.

These cataclysmic events live on in Indian myths, such as that which the indigenous people tell about the formation of the Grand Canyon through a huge flood. As described at the beginning of the book, this scenario has now been scientifically proven, as parts of the Grand Canyon only formed 1,300 years ago. Was the Sierra Nevada region in California not also ploughed up by floodwaters and volcanic eruptions? This region lies west of a volcano belt, which runs in a north-south direction and extends as far as Alaska.

Precisely this region of the Sierra Nevada lies in the area of a very active hot spot, which may be compared with that in the area of the East African Rift Valley. Intense volcanic activity caused the destruction of flora and fauna in East Africa and in the West of America, leaving behind a crushed mixture of lava and andesite boulders, completely saturated by mighty masses of water. Immediately afterwards, these flooded landscapes with several layers were cut apart by torrential rivers and covered with lava, which hardened into huge basalt domes. Thus it is not pure chance that seemingly very old objects are found in the area of the Sierra Nevada.

Further north, not far from Seattle, the volcano Mount St. Helens stands in the area of the Cascade Range, a mountain range with many volcanoes (including Glacier Peak and Mount Rainier) which runs along the Pacific Coast in a north–south direction from British Columbia to California. Lassen Peak in north California, which belongs to the 50 to 80 km wide volcano belt, erupted in 1911.

The effects of the Mount St. Helens eruptions in 1980 and 1983 were gigantic and were geologically studied and examined in-depth. In 1980 this flourishing area, rich in trees, was transformed into a lunar landscape in a matter of hours; table mountains of up to 50 metres in height were formed, into which a mix of tree trunks, house ruins, parts of cars and technical appliances were swirled. This caused the mixing of old and new volcanic material and the respective artefacts therein.

Masses of water poured over the deposited mud layers and shaved their upper surfaces smooth. Mud layers of up to 50 metres in height and more were formed, which looked as if they had been levelled off with a palette knife on top. The water masses emitted by the volcano then ate into these fresh mud layers. New river valleys were formed and these, in turn, separated the mud layers into separate, isolated areas: table mountains were formed. In only a few hours, a green, forested landscape was transformed into a desolate mud landscape. Artefacts buried in these new table mountains and exposed in the newly–formed river valleys are the same age!

Our geologists were able to study this scenario of a sudden landscape transformation in detail, although only very few experts were really interested or even visited the site. In the Sierra Nevada when a similar scenario took place a few thousand years ago, there were no geologists present but only indigenous peoples or Neolithic settlers who preserved these events in their myths, which were then casually rejected by the geologists as pure fantasy.

This attitude once again shows how the blinkers prescribed by Science block sight of the most valuable clues.

The geological observations which I have continually undertaken since 1987 with regard to the superfloods in this area, have been confirmed by a new study. In the southern region of the Sierra Nevada proof of alluvial sedimentation caused by flash floods from the Sierra Nevada mountains, in the area of Mount Whitney and Lone Pine City has been found (Blair, 2002, pp. 113–140).

Since Tertiary layers are completely absent in this flood region, the flood layers, deposited in several separate layers one on top of the other, lie directly on top of volcanic plutonic rock, rich in silica, which penetrated the older volcanic Jura layers in the younger Cretaceous Period (ibid. p. 117; cf. Evernden/Kistler, 1970).

These flood layers are officially categorised as Ice Age end moraines, i.e. the till and gravel of melting glacier tongues.

If this scenario described in connection with Mount St. Helens and the Sierra Nevada is correct, then the geological age determinations are definitely wrong. Since the (Tertiary) layers of the Sierra Nevada which have allegedly been growing ever-so slowly for 55 million years over the existing rock (primary rock) were actually formed cataclysmically in a short space of time (time–impact), one can logically reduce the millions of years which would be needed for the slow formation of Tertiary layers to only a few years at most. Conclusion: the Tertiary Period of the Sierra Nevada is a phantom period. Otherwise it would appear, if the official geological chronology is to be believed, that the existence of humans countless millions of years would thus be proven.

Let's now take a look at another artefact from the Tuolomne table mountain which was discovered in situ by Clarence King, then one of America's most famous and highly respected geologists. In compact, hard, auriferous gravel, a piece of a cylindrical stone pestle was found wedged into the hard gravel. After it was removed, a perfect impression was left in the rock (Becker, 1891, p. 193f.)

The facts described rule out a secondary deposition. Even Holmes who had criticised the Tuolomne finds (1899, p. 453) conceded after examining the site, that this *find may not be challenged with impunity* even though he registered the presence of some modern Indian millstones in the vicinity. The King pestle now forms part of the Smithsonian Institution's collection. Modern geologists discredit this find as it was excavated from a layer which was allegedly 9 million years old, pointing out that the matrix with the impression of the pestle can now no longer be verified, even though the find itself and the published report are both available.

This argument is absolutely absurd since for almost all officially recognised paleo-anthropological discoveries it is exclusively finds and reports which provide the sole legitimation of a find. If certain discoveries fit into the plan of things and into the sacrosanct time structure of geology and evolution, then a missing matrix is, of course, no problem for the experts. In fact, there is hardly any other fossil find for which this type of "negative print" is available.

The most famous fossil was discovered in February 1866, at a depth of 40 metres, in a mine on Bald Hill close to Angel's Creek in California: the famous Calaveras Skull, the age of which was said to be 55–33 million years (Whitney, 1880, pp. 267–273; cf. *Handbook of American Indians*, I, p. 188f. and Schmidt, 1894, p. 31ff.).

Up to today it is still disputed whether this skull is authentic or, coming from a young Indian burial site, a fake. I don't intend to discuss this case here, although Sir Arthur Keith said: The story of the Calaveras Skull cannot be passed over. It is the "bogey" which haunts the student of early man taxing the powers of belief of every expert almost to breaking point (Keith, 1928, p. 471). After all, Professor Wyman, who was to scientifically describe the find, had great difficulty in removing the cement–like gravel with which the skull had fused (Journal of Transactions of the Victoria Institute, 1880–1881, pp. 191–220).

But the Calaveras Skull does not represent a single, isolated discovery. A skull fragment, named after Paul K. Hubbs, was found at a depth of 25 metres in auriferous till close to a jumble of mastodon bones. The find site lay beneath a compact and hard basalt dome, which bears witness to volcanic activity in the area at the time. This bone fragment is now in the museum collection of the Natural History Society in Boston and is labelled as follows: "Found in July 1857. Given to Rev. C.F. Winslow by Hon. Paul K. Hubbs, August 1857".

A similarly labelled fragment from the same skull was also in the museum of the Philadelphia Academy of Natural Sciences. It is clear that we would never have heard of this skull fragment, had Mr. Hubbs not been on the spot, as the fragment would never have been found (Whitney, 1880, p. 265).

According to Whitney, all human fossils which were found in the gold–mining region were of an anatomically modern type. These included a human jawbone from the Tuolomne table mountain (Whitney, 1880, p. 264ff.) and other bones which were discovered from 1855 to 1856. In another gallery, on the same level at which mastodon teeth and elephant bones were found, the entire skeleton of a modern human was excavated (Winslow, 1873, pp. 257–259).

Given that the mastodon teeth were found close to the primary bedrock, one must assume going by geological chronology, that with an age of 55–33 million years old, humans and mastodons are many times older than the official evolutionary ladder allows.

Are these geological age determinations correct? Mastodons lived in America until the end of the "Ice Age" up to 12,000–10,000 years ago. Until the "Ice Age" America's vast plains, including the Sierra Nevada in California, resembled today's African Serengeti. It is accepted that huge herds of bison grazed here, alongside wild camels, llamas and wild horses. Gigantic woolly mammoths grazed beside herds of mastodon. There were beavers as big as bears which, in turn, were even more gigantic, bigger than any polar bear alive today. Apart from various small sloth species there was also a giant sloth the size of an elephant.

American lions were bigger than their modern day African counterparts or the legendary sabre–tooth tiger. There is also evidence for the co–existence of humans, giant sloth and mastodon. In May 1839, Dr. Albert C. Kochs discovered carbonised mastodon bones together with stone axes and arrowheads along the Mississippi in Missouri. However, not only did humans live alongside these primeval giant mammals, but seemingly also with animals who are supposed to have lived in the late Tertiary Period (Pliocene) well over 2 million years ago.

Anomalously Old Tools

On 25 November 1875 the Professor of Geology, Giovanni Capellini of the University of Bologna reported a sensational discovery, which he had made while cleaning the completely fossilised bone of an extinct small whale: a sharply cut notch. This must have been made with a sharp instrument before fossilisation because it is impossible, even with a steel blade, to scratch the surface of the fossilised bone (de Mortillet, 1883, p. 56; cf. Cremo/Thompson, 1997, p. 75).

Further examination revealed three more notches. It appears that the notches were made by a very early human since the bones can from the extinct small whale Balaenotus, which was typical of the younger (Upper) Pliocene in Europe over 2 million years ago. The whale bones were probably deposited in the shallow coastal waters before the old island of Monte Vaso in the region of today's Tuscany. At the end of the Tertiary this island belonged to an archipelago in the region of today's central Italy.

To test his hypothesis of man–made notches, Capellini examined the bones of freshly–slaughtered animals. They showed a similar pattern of cuts. He then tested old flint tools which he had excavated in the region. It was seen that the same type of notching could be made in fresh bones.

To his astonishment, Prof. Capellini found that traces of cuts were visible on a large number of bones but only on the upper part of the spine and on the outside of the ribs on the right side. Because of this specific distribution, Capellini developed the theory that the whale had been stranded in the shallow water of the primeval sea and had been lying on his left side. Humans then cut the flesh out of his right side using flint tools.

Capellini saw this as proof that human must have lived in Tuscany at the same time as the fragmented whale over 2 million years ago. He presented his findings in 1876 in Budapest and in 1878 in Vienna at international congresses (Capellini, 1877). His findings were confirmed by several other scientists (Binford, 1981, p. 111).

The museum in Florence received a large collection of whale bones which had been excavated in the Fine Valley in Tuscany. Capellini found identical, obviously man–made cut marks on these bones too. Further similar bones appeared in Italy. A fossilised animal bone from either elephant or rhinoceros displayed a round, drilled hole at the centre of its widest point. This bone, which had been firmly embedded in an Astian (Upper Pliocene) layer over 2 million years old in San Valentino, was presented at an 1876 conference of the Italian Geological Committee, "with traces of human working which were so obvious that all doubts were removed" (de Mortillet, 1883, p. 73).

Was this a locally isolated phenomenon? No: in seashell deposits at Barrière near Pouancé in northwest France, Abbé Delaunay discovered the bone of a Halitherium, an extinct manatee, which bore notched markings which were obviously man–made. The find caused a sensation when it was presented to the members of the International Congress for Prehistoric Anthropology and Archaeology in Paris in 1867 (de Mortillet, 1893, p. 53).

The fossilised bone was firmly embedded in an undisturbed layer. This maritime deposit is now dated to the older (Lower) Miocene. Humans, who obviously fragmented this Halitherium, must have lived around 20 million years ago or even earlier. "That is far too early for humans", wrote de Mortillet (1883, p. 55). If advocates of evolution theory express themselves in this manner, their prejudice in interpreting facts and finds is clearly demonstrated. De Mortillet interpreted the notches on the bones as shark bite marks.

In the book Human Origins, this question is discussed in detail: "The cuts can be compared with those on thousands of undoubted human cuts on bones from the reindeer and other later periods, and with cuts now made with old flint knives on fresh bones .... even an ordinary carpenter would have no difficulty in distinguishing between a clean cut made by a sharp knife and a groove cut by repeated strokes of a narrow chisel" (Laing, 1894, p. 353f.).

A study of the man–made cuts on bones from the Olduvai Gorge reached the following conclusion: The scratching and gnawing of bones by carnivores leaves grooves with a round or flat surface; in both instances the fine, parallel grazing of cut or scratch traces is absent (Potts/Shipman, Nature, 1981, p. 577). And a modern expert writes: "It is rather unlikely that one ... could confuse cut markings created during the dismembering or deboning of an animal's bones using tools, with the marks left by carnivore activity" (Binford, 1981, p. 169).

There have been a great many other finds of cut marks on fossilised bones which are far too old to fit into the propagated history of mankind. To give another example, an allegedly 15 million year old upper thigh bone of an extinct rhinoceros was found at

Fig. 23: Shelled, fossilised Sea Urchin. *The insides have been pushed up to the highest point (arrow) and damaged. A knife may have entered the sea urchin and, after the shell was broken open, was pushed in exactly to the depth of the shell. The cut was from above to the front to the underside of the mouth and ended at the mouth. The other end of the circular cut comes to the other side of the mouth, as orientation point, through which the cut was supposed to run despite diversion from the symmetrical axis. Photos and interpretation: Volker Ritters, 1998.*

Gannat, France, whose surface displays parallel cut marks (de Mortillet, 1883, p. 52).

According to de Mortillet, the short parallel grooves were caused by subterranean pressure, in other words they are the marks of purely geological phenomena. According to Binford however, (1981, p. 169), these are specific markings, unmistakably caused during slaughter.

A fossil sea urchin, half–peeled and showing its soft parts cannot exist because the exposed insides would not have remained fresh and well–preserved long enough to have fossilised slowly. Yet Volker Ritters describes just such a sea urchin which is in his possession and this presents another riddle: the shell was completely bisected by a clean cut (Ritters, 1998, p. 7). But this fossil sea urchin (*Ananchytes ovata*) was one of the most common species of the Upper Cretaceous. Who was peeling sea urchins with a sharp tool during the earth's middle period 70 million years ago at the time of *Tyrannosaurus Rex?*

If humans were peeling sea urchins, slaughtering animals and leaving cutmarks in bones x–million years ago, then there should be correspondingly old tools. A look at the literature shows that in Europe in many rock layers, which are allegedly too old for such artefacts, advanced stone tools have indeed shown up.

In 1899, the French archaeologist, Eugène Bonifay discovered at Saint-Eble in central France a number of simple stone objects in a layer of volcanic ash. The volcano which is now extinct is said to have erupted 2 million years ago (*Science*, Vol. 246, 6 Oct 1989, pp. 28–30).

Over a hundred years ago in Ightham, England, stone tools more advanced than the objects described above, were found in layers which are 4–2 million years old. They reveal a remarkable similarity with those tools excavated by the Leakeys from the Olduvai Gorge.

Numerous shark's teeth from a Charcharodon, said to be 2.5–2 million years old have been found in a Red Crag formation. Numerous examples were presented in 1872 at an anthropology conference. They were all bored through in the middle, as is seen amongst the South Sea Islanders who make weapons and necklaces out of them (Nilsson, 1983, p. 106). Ascribing the boring to mechanical human activity appeared the most likely explanation of the facts (Charlesworth, 1873, p. 91f.). Natural wear and tear or tooth decay would hardly leave clean round holes in the middle of a series of teeth.

A large number of worked flintstones, scrapers and hand axes, were also found in situ in this Red Crag formation in several places, deep below the surface. An international commission of prehistory experts was set up to examine the artificially worked flintstones from the deepest layer of the red crag near Ipswich. In 1923 the commission ruled "There exist at the base of the Crag, in undisturbed strata, worked flints (we have observed them ourselves). These are not made by anything other than a human or hominid which existed in the Tertiary epoch. This fact is found by American prehistorians to be absolutely demonstrated (Lohest et al., 1923, p. 67).

Thus it was scientifically proven that man–made tools had been dated to an age of 2–5 million years. This result needs to be double-underlined. It is confirmed that during Lucy's lifetime, tools were already being produced and used in England and Ireland – the red crag regions. Were there simultaneous developments in Europe and Africa? Unlike in Africa with Australopithecus, however, in Europe, there were apparently no ape-like human forerunners.

Doesn't this mean that the commission's findings need to be revised? Yes! But today these, like most of the cases documented in this book, are forgotten, because the army of progress objectors – the scientists of earth and human science and the mass media – have scarcely any interest in questioning evolution theory: the money's too good and there are almost no doubts. This is why scientifically acknowledged anomalously old finds are today forgotten and unknown.

If anybody mentions the 1923 results and thus the existence of Tertiary man, this is rejected as an error because it contradicts the theories arising out of the African excavations. If only one of the finds described here is authentic and the geological dating is correct, then evolution theory would be pulverised and revealed as the greatest error of the millennium.

In the history of paleo-anthropology there are a few other papers, analogous to the previously described discoveries, which deal a death blow to prevailing views. Of course, such finds of flint tools (eoliths) in anomalously old layers cannot remain uncontested. Abbé Henri Breuil (1910) already wrote a "work of destruction" at the beginning of the

20th Century. He was writing about finds from a gravel pit at Clermont, north of Paris.

A clay layer with tiers of square–edged flints intermixes with greenish Bracheux sand, belonging to the Lower Eocene layer beneath a chalk bed (Obermaier, 1924, p. 12). Were these flints (eoliths) worked, by chronological reckoning, by humans as early as 50 million years ago? In order to answer this question in the negative, of necessity these tools must be confirmed to be of natural and not man-made origin. In Breuil's opinion: these are flints which were inside the bed when the flaking took place, whereby the fragments remained in close contact with each other" (Breuil, 1910). Such natural flaking does indeed occur, but "pressure flaking very rarely produces clearly marked bulbs of percussion. It usually takes an intentionally directed blow (*Journal of Field Archaeology*, Vol. 10, 1983, pp. 297–307).

If Breuil is correct in his assumption that prominent edge retouching is caused by geological pressure then none of the officially recognised finds showing traces of rough percussion, including those in more recent layers, should be accepted as proof of human activity!

Accordingly, the largely rudimentary Olduvai stone tools would have to be rejected as scientifically worthless as well. Since the 50 million year old eoliths from Clermont resemble those produced by *Homo erectus* (Acheulean) which are 200,000–1.5 million years old, many other stone tools from these Old Stone Age periods must also be said to have been produced by accident. This would, however, drastically reduce the already small number of eoliths found. In this case, was there a Stone Age at all? The proponents of eoliths make reference to the Australian Aborigines, who still produce similar tools nowadays.

The rejection of 50 million year old eoliths also stems from evolution theory – it is a matter of faith: "From a paleo-anthropological point of view this is all untenable. The closest relatives of the Eocene man of Clermont were Pachylemurs (an extinct genus of lemur)!" (Schlosser, 1911, p. 58; cf. Obermaier, 1924, p. 16f.).

Like their modern day counterparts, experts such as Breuil (1910), Schlosser (1911) and Obermaier (1916) argued that if the stone tools were man–made, there must also be skeletons found in the (for evolution theory) anomalously old layers as well. This argument illustrates the more than questionable methods of paleo-anthropology because none of the finds described in this book of fully human skeletal remains from the Pliocene, Myocene, Eocene and other, even older geological periods, which must be classified as modern humans have been officially accepted even when they have been discovered by experts.

By rejecting finds of Tertiary humans and the corresponding eoliths in practically all Tertiary layers since the dinosaur era because they are out of step with evolution theory on the one hand, and claiming that the alleged find scarcity in these layers explicitly proves the rightness of evolution theory on the other, paleo-anthropology reveals its questionable reasoning to be untenable and illogical – a proof that proves itself.

An Ancient Hut

It appears that a building technique which has been around for only a few thousand years, was already known when, an alleged 2 million years ago, *Homo sapiens* started to develop in the shape of *Homo erectus*.

In the 1960s, Louis Leakey made a sensational find in the Olduvai Gorge (northern Tanzania). In the Bed II Layer, he discovered that Australopithecus, *Homo habilis* and *Homo erectus* coexisted.

Alan Walker confirms this fact when he says: There is evidence from East Africa for late–surviving small Australopithecus that were contemporaneous first with H. habilis, then with H. erectus (*Science*, Vol. 207, 1980, p. 1103). In the same layer (Bed II), Leakey however also discovered the remains of a stone hut. The aspect of this find which caused a stir was that the type of construction, which is still seen in some parts of Africa, can only have been carried out by *Homo sapiens*. On the basis of Leakey's discoveries, Australopithecus, *Homo habilis*, *Homo erectus* and modern humans lived contemporaneously around 1.7 million years ago (Leakey, 1971, p. 272 and Kelso, 1970, p. 211).

The palaeontologist, Stephen Jay Gould of *Harvard University*, a well-known evolutionist, explains evolution's dead end as follows: What has become of our ladder if there are three coexisting lineages of hominids (A. africanus, the robust australopithecines and H. habilis), none clearly derived from another? Moreover, none of the three display any evolutionary trends during their tenure on earth (*Natural History*, Vol. 85, 1976, p. 30). One must agree with Gould without reservation; he exactly pinpoints evolution theory's unsolvable dilemma.

5   The Neanderthal Falsity

The paleoanthropologist, Dr. David Pilbeam, Professor at Yale University asserts: Perhaps generations of students of human evolution...have been flailing about in the dark; that our data base is too sparse, too slippery, for it to be able to mold our theories. Rather the theories are more statements about us and ideology than about the past. Palaeontology reveals more about how humans view themselves than it does about how humans came about, but that is heresy (American Scientist, Vol. 66, May/June 1978, p. 379) – or the truth, for conventional teachings are simply wrong.

The Evolution of the Neanderthals

The palaeontology elite is still in disagreement with regard to the basic outline of the human family tree. New branches burst into bloom with great pomp and ceremony, only to wither away again and die when new fossils appear. Without attracting much attention, Neanderthal representations of bowed, ape-like beings were replaced with more human, upright models in natural history museums all over the world.

If Neanderthal Man could be reincarnated and placed in a New York subway – provided that he were bathed, shaved, and dressed in modern clothing – it is doubtful whether he would attract any more attention that some its other denizens (Strauss/Cave in *Quarterly Review of Biology*, Vol. 32, 1957, pp. 348–363).

The physical features of the Neanderthal which were previously deemed primitive on the basis of Marcellin Boule's reconstruction (Ann. Paléontol., 7/1912, pp. 105–192) were regarded as an expression and result of a lower cultural stage (Stringer/Gamble, 1933; Trinkaus, 1983) and a simpler social organisation (Trinkaus in Journal of Human Evolution, Vol. 25, 1993, pp. 393–416). This was also seen as the reason for the Neanderthal becoming extinct: according to Darwin's law of the stronger (now known as survival of the fittest) he would have had to have made way for the seemingly higher–developed modern human.

This brutish and primitive image of a bowed creature with bent knees and bestial behaviour was upheld for over 50 years to provide a visualised picture of human evolution and, nourished by the conspiracy of media and science, became dogma. This systematically applied ideology – or brainwashing – worked well for decades: human evolution was literally planted in the sub–conscious and took root there. However, the history of human evolution was invented to satisfy interests other than the solely scientific. But the Neanderthal has been undergoing a real "evolution" for the last few years. After all, the Neanderthal's brain volume is greater than that of the average

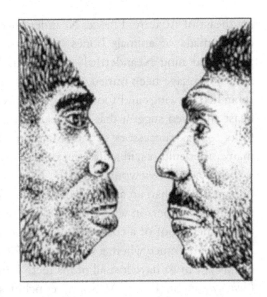

Fig. 24: **Neanderthal.** *Using one and the same skull, a reconstruction with modern or ape-like profile can be created, depending on the individual position of the anthropologist. After Junker, 2002, p.26.*

modern human. The cranial capacity of a Neanderthal skeleton found in Israel's Amud Grotto was, believe it or not, 1,740 cubic centimetre! Today's humans have an average volume of just over 1,400 cubic centimetres and Albert Einstein's brain was actually 12% less than that. Although one cannot extrapolate a direct link between brain size and intelligence, R.L. Holloway discussed the Neanderthal brain – which was by no means primitive – in the *American Journal of Physical Anthropology Supplement* (Vol. 12, 1991, p. 94).

Detailed comparisons of Neanderthal skeletal remains with those of modern humans have shown that there is nothing in Neanderthal anatomy that conclusively indicates locomotor, manipulative, intellectual, or linguistic abilities inferior to those of modern humans (Erik Trinkaus: *National History*, Vol. 87, Dec. 1978, p. 10).

The Neanderthals apparently also possessed the technical means to produce sophisticated tools and a type of superglue. Finds in Germany's North Harz region show that the Neanderthal produced a pitch from birch bark in order to glue stone blades to wooden shafts. To produce this pitch, however, a constant temperature of 360°–400°C has to be maintained for a long period. This adhesive, which was seemingly also used as chewing gum (*BdW*, 16 Jan 2002), cannot be discovered or produced by sheer chance (*BdW*, 8 Jan 2002).

Skilled artisans also require fine mechanical skills and hands which are not apelike. A computer analysis showed that, unlike the great apes, the Neanderthal could form an "O" with his thumb and forefinger with no difficulty. Scientists stress that this is a prerequisite for working with tools. Furthermore, the hands of our extinct cousins were as flexible as those of modern humans (*Nature*, Vol. 422, 27 Mar 2003, p. 395).

We know that Neanderthals also had a predilection for music, since they produced it using the diatonic scale. Irregularly spaced holes in a flute made of bear bone, found in Slovenia in 1995 indicate that semitones and full tones could apparently already be produced 50,000 years ago (*Scientific American*, Sept. 1997).

Jewellery was also produced at an early stage. Fragments of beads made from ostrich egg shell said to be 200,000 years old were found in El Greifa, Libya. In Arcy–sur–Cure southeast of Auxerre, France, Neanderthals ornamented themselves with ivory rings and chains made of animals bones and teeth. Between 1953 and 1960, Ralph S. Solecki discovered nine Neanderthal skeletons in the Shanidar Caves in Iraq. One of them appears to have been buried with flowers as a grave good. Grave goods such as flowers, scratched drawings and food, completely overturned earlier impressions of a half–human beast and even suggest that the Neanderthal believed in an afterlife (Solecki, 1971). The Celts too, as successors of Cro-Magnon Man and the megalithic culture believed, like many other cultures, in a life after death.

From a social viewpoint too, the Neanderthal appears to have behaved like a modern human rather than an ape, because one of the skeletons found in the Iraqi cave suggests that the living person was partially blind, one-armed and crippled. That he survived at all is a further proof of a complex social structure. This is also confirmed by another find in L'Aubesier, France where a supposedly 179,000 toothless jawbone was found. The owner of this jaw must have lost all of his teeth long before he died. Erik Trinkaus of *Washington University* in St. Louis (Missouri) concludes that this toothless person must have had special food prepared for him by members of his group, over a certain period of time (*Journal of Human Evolution*, November 2002; cf. *Science*, Vol. 301, 5 Sept 2003, p. 1319). It seems, therefore, that at that time there were already clans with a corresponding social structure.

In *Science* (Vol. 299, 7 Mar 2003, pp. 1525–1527), it is confirmed that there were no differences in behaviour between Neanderthals and early–modern Cro–Magnon Man in terms of tool manufacture, burial, interest in mineral pigments, control of fire and dependency on meat supply. Furthermore, the skeletons of both species reveal physical weaknesses and handicaps which would have required care by clan members. The cultural, social and humanitarian consonance that exists between these two Stone Age peoples can hardly be closer (Klein, 2003).

The long-held assumption that Neanderthals were incapable of modulable speech has also been proven invalid: in 1993 the discovery of a hyoid in the Kebara Cave in Israel showed that the structure of the voice box in Neanderthals and modern humans is identical. Some scientists still believe that speech only came to Europe 40,000 years ago with Cro-Magnon Man but this opinion was contested by scientists at *Duke University* in April 1988: the Neanderthal was capable of speech (*BdW*, 17 Feb 1999). Even the differences in body structure are far less significant "if one does not define the Neanderthal based on extreme finds" (*BdW*, 1 Oct 1996).

Conclusion: The Neanderthal barely differs from modern Man, (*Nature*, Vol. 394, 20 Aug 1998, pp. 719–721). One of the leading Neanderthal researchers, Erik Trinkaus concludes that: Detailed comparisons of Neanderthal skeletal remains with those of modern humans have shown that no conclusions with regard to difference can be drawn in anatomical terms with regard to locomotor abilities, intelligence and speech (*Natural History*, Vol. 87, 1978, p. 10).

Species or Sub-Species

For decades, the Neanderthal was classified as an individual species (Homo neanderthalensis). This made it clear that our "forerunner" was more primitive than modern Man and could not interbreed with him. Once it was recognised that Neanderthals and modern Man are the same apart from a somewhat different body shape, the Neanderthal was demoted to a human sub–species. Some scientists then dubbed him *Homo sapiens* neanderthalensis.

However, genetic studies by a group of scientists and the anthropologist, Svante Pääbo, in July 1997, produced a spectacular finding. The specialist journal Cell which published the findings headlined the article: "Neanderthals were not our ancestors", even though the scientists had expressed themselves more carefully.

The skulls of 225 living people were compared with five fossil samples of *Homo sapiens* and with five Neanderthal skulls and, up to now, this is supposed to be the most solid proof "that the Neanderthal is in fact a distinct species within the Homo genus" (*PNAS*, 3 Feb 2004, Vol. 101, pp. 1147–1152). One might dispute this finding, especially when one sees the varying skull forms of modern humans. Within a given species, however, isolated populations which differ morphologically, physiologically or in some other way from other populations, but are still part of that species, may develop over only a few generations.

In the event that the Neanderthal is finally proven to be a separate species, it is assumed that he shared a common ancestor with modern Man 600,000–500,000 years ago (*Nature Review Genetics*, Vol. 2, 2001, p. 353) who, according to Tattersall (1995) may have been Homo heidelbergensis (formerly known as archaic *Homo sapiens*).

If this proves to be the case, where was modern Man hiding all that time? The oldest *Homo sapiens sapiens* is said to have lived 140,000 years ago and, in Africa 200,000 years ago going by new finds in Ethiopia (Omo 1 and 2). But there is no proof until 30,000 years ago, so that 100,000 years are unaccounted for. It is said that early–modern Man arrived in Israel 100,000 years ago but here too, there is no continuous presence in evidence until 30,000 years ago. What use is an old find when practically no further finds are made from the time when the split from the forerunner, supposedly Homo heidelbergensis was made? If the Neanderthal is a separate species, the Neanderthal chain of development which must also be proven, is absent.

If one goes by the empirical study published in *Nature Genetics* (Vol. 15, 1 April 1997, pp. 363–368) and shortens the calculated time at which the common ancestor of Neanderthal and modern Man was alive by a factor of 20, he would then have lived only 30,000 years ago, rather than 600,000 years.

Many scientists currently assume, however, that the Neanderthal was a sub–species of modern Man. There may therefore also have been mixed race children of both Neanderthal and modern Man origin, because the Neanderthal's competitor was supposed to have been Cro-Magnon Man, an early-modern human. This type was named after the cave in the Dordogne where the first fossil remains were found. Modern Man supposedly came from Cro-Magnon Man but this presupposes a development which never took place because Cro-Magnon Man is identical to modern humans. There is no evolution. Naming Cro-Magnon Man after a cave only serves to conceal this fact.

Did They Just Disappear?

For well over one hundred years, we have been led to believe that early modern Man (Cro-Magnon Man) displaced and exterminated the "apelike" Neanderthal wherever he encountered him. With the findings mentioned above, this opinion has started to change. The paleoanthropologist, Ralph L. Holloway of Columbia University in New York says that all brain asymmetries which characterise modern humans are also to be found in Neanderthals: To be able to discriminate between the two is, at the moment, impossible (Wong, 2004, p. 71). Why then did the Neanderthal die out, if he was at least the equal of early–modern Man in terms of intellect and manual dexterity?

For a long time scientists were convinced that the Neanderthal was pushed out to the Iberian Peninsula by the westward expansion of Cro-Magnon clans, because the oldest remains, said to be 32,000 years old, were found Zafarraya in Spain. Recently, however, finds in Vindija, Croatia have been radiocarbon dated to 29,000–28,000 years old (*PNAS*, Vol. 96, 26 Oct 999, pp. 12281–12286). Vindija is however in the middle of a region which was supposedly already ethnically "cleansed" in an earlier phase of early–modern Man's expansion. For the advocates of displacement, this proof of Neanderthals' late existence is a bitter blow. On the other hand, this find makes the Neanderthal even younger and the phase of coexistence longer, if the dating is correct.

There is a scientific alternative to the displacement theory: hybridisation theory. "Neanderthals developed into *Homo sapiens*" says Milford Wolpoff, an anthropologist at the University of Michigan. The basis for this assumption was the Vindija find. The new dating supposedly shows that both types of human co–existed in Europe for several thousand years, as is supposedly also proven by the find of an early–modern jawbone, 34,000–36,000 years old, in the Pestera cu Oase Cave in Romania. Thus they would have had adequate time and opportunity to mix biologically.

The almost complete skeleton of a four year old child in the central Portuguese Lapedo Valley allegedly provides further support for this hybridisation theory. Since the bones displayed characteristics of both types of human, they are proof that interbreeding may have taken place between Neanderthals and *Homo sapiens* (*PNAS*, Vol. 96, 22 June. 1999, pp. 7604–7609). The bones have been controversially dated as 24,500 years old. This is why other scientists are convinced that this must have been a modern human, given that Neanderthals were supposedly already extinct (*Science*, 30 April 1999, p. 737).

Ian Tattersall of the American Museum of Natural History in New York said in an accompanying commentary on the mixed–race child that its body proportions corresponded with those of an early–modern Man adapted to a cool climate. This is a feature which we will be looking at in more detail.

Even if there were mixed–race children, new genetic analyses of the remains of 24 Neanderthals and 40 early–modern humans show that there was no mixing of the two species (*PNAS*, 1 Oct 2002, Vol. 99, pp. 13342–13347). Since not all Neanderthal genes were examined, it is possible that modern Man received a few genes from him (*Science*, Vol. 299, 7 Mar 2003, pp. 1525–1527). Since there were few such sexual contacts, these have not left significant genetic traces in the long term. It appears on the basis of bone dating, that Neanderthals and early–modern humans lived alongside each other for thousands of years. But do age determinations provide reliable results? Richard G. Klein says that samples said to be 50,000–40,000 years old may be 20,000–10,000 younger because of contamination of the carbon examined. Journals such as *Science* (Vol. 299, 7 Mar 2003, pp. 1525–1527) have recently also started to question the error-proneness of dating methods. As has already been seen, age estimates are often undertaken arbitrarily, or even simply invented!

Generally speaking, age cannot be determined by using direct dating methods. Most find sites do not contain bones, but rather a certain type of stone tool, from whose method of production the presence of Neanderthals or modern humans is inferred. Is this not simply a record of a technique which both groups may have used? In addition, stone tools cannot be dated directly, as there are no measuring methods to do so. Thus it is not so easy to assign certain finds to specific periods – in fact; it can be done only indirectly or not at all. However, when experts work on the basis of a recorded and fixed chronological timescale based on geological layers, and then we start to see paradoxes.

Some scientists come to the peculiar conclusion "that only geological layers got mixed up", since artefacts of early-modern humans (Aurignacians, starting 40,000 years ago in Europe, at most) often came to lie beside those of late Neanderthals (Châtelperronian: around 34,000–30,000 years ago). "Other scientists believe that such things come from modern Man. The Neanderthal either picked them up, bartered them or imitated their style of manufacture, without really understanding the symbolic significance of the objects" (Zilhão/d'Errico, 2004, p. 68). Again, is this not simply the documentation of a transitional phase, in which both techniques of both groups were used until the better method asserted itself throughout a given region? João Zilhão (Portuguese Institute of Archaeology, Lisbon) and Francesco d'Errico (*University of Bordeaux*) re–evaluated the material from the Grotte du Renne and came to the conclusion that these fossils and artefacts from different cultural epochs actually belonged together because complete objects and waste are lying in the same layer (Zilhão/d'Errico, 2004, p. 68). What for a long time seemed to separate moderns from Neanderthals – the ability to produce symbolic cultures – has definitely collapsed". (Zilhão/d'Errico, 2004, p. 69).

Federico Bernáldez de Quirós and Victoria Cabrera who already worked at the El Castillo site excavated by Obermaier between 1910 and 1915, "see no differences between the economic structure or lifestyle of the Moustérian inhabitants of the cave (Neanderthals) and the inhabitants of the Aurignacian layers (Cro-Magnon Man), which lie immediately above and have been dated to around 40,000 years. They also see a high degree of continuity with regard to the stone tools. Can this be a single species of human?" (Arsuage, 2003, p. 310). This would then be a record not of displacement but of continuity.

The conclusion reached by these examinations appears to be correct: the artefacts assigned to the Châtelperronian (Neanderthal) culture superficially resemble early–modern Aurignacian artefacts, simply because they were produced using a different, older technique. It is therefore no mystery why artefacts are in the same layer. The error must be in the circumstance that conventional teaching always associates certain artefacts or decorative patterns with one particular culture. Thus fictitious Châtelperronian, Aurignacian or band–ceramic peoples are created, although what we are looking at are really only cultural and technical stages. Since there are usually no fossilised human bones found, but only stone artefacts produced using a specific technique, the presence of certain and differing human species is inferred.

Accordingly, one could infer the existence of three difference peoples in Germany in the 20th Century. Before the First World War, during the inter–war years and after the Second World War, seemingly completely disparate cultures, who differed entirely from one another in terms of art, architecture, technology and systems of government, became native. Thus can the illusion of a mass migration or even the extinction of certain ethnic groups be falsely created. Similarly, based on the worldwide distribution of Coca Cola cans, one might assume that there had been a global migration of Americans after the Second World War. When no artefacts which were produced using an (allegedly) older, Neanderthal-ascribed technique are found in younger layers, it is assumed that the Neanderthals must have become extinct. Could it not simply be that an older technique was dropped and a technical innovation introduced? Is the fact that steam engines are now only found in museums proof that the people of 19th Century Europe became extinct?

Moreover, our ancestors already lived in tents and houses, and not in caves, as has been proven by many finds. According to official dating, Europe's oldest hut was already standing around 600,000 years ago in Prezletice (East Prague) in the Czech Republic. Germany's oldest huts were found during excavations in Bilzigleben, Thuringia and are said to be 300,000 years old. In the Old Stone Age there was also long–distance trading and thus peaceful cultural exchange and the transmission of new techniques. Did Neanderthals and early–modern Man not simply just use the same techniques? The Neanderthal expert, Prof. Gerhard Bosinski, head of the *Neuwied Museum* for Ice Age Archaeology emphasises: "As we know today, the Neanderthal's lifestyle did not vary

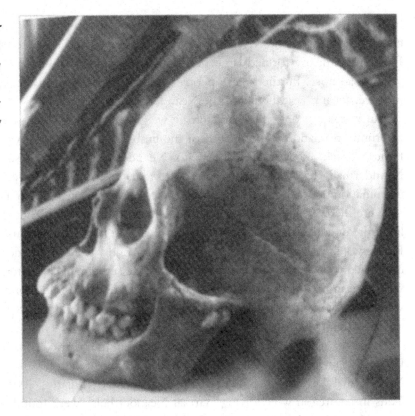

Fig. 25: Other Indians. A "flathead" from Crosby County around 1600. Compare photo 53: skull from the Kow Swamps in Australia.

fundamentally from that of modern Man" (*BdW*, 1 Oct 1996). The Châtelperronian skeleton of a Neanderthal found at St. Césaire "differs little from early Neanderthals and gives no indication of any evolution into modern Man" (Gambier, 1989, p. 207). Was there no Neanderthal development in Europe and is this why the Neanderthals became extinct?

They're Still Alive

Following the discovery of the first Neanderthal in 1856, various newspapers including the German Newspaper *Bonner Zeitung* printed a short report stating that during the clearing of clay mud a human skeleton had been found, which obviously "belonged to the race of Flatheads". This race, ran the report, still populates the American West and the question arose as to whether the skeleton's owner was a member of a primeval European people or whether he had been one of Attila the Hun's roving horde. In *Nature* (Vol. 85, 8 Dec 1910, p. 176) reference was made to an article in the *Philippines Science Magazine* (June 1910) written by Dr. R. B. Nean of the Anatomical Laboratory in Manila. It was reported that on the Philippine island of Luzon, a group of Old Stone Age people had been discovered. These individuals possessed similar skull characteristics to the classic flat–browed Spy Neanderthal, (named after the place where he was discovered in Belgium): similarly formed large heads with prominent browline, heavy lower jaw and very broad nose. Further, these people were around 160cm in height and displayed the massive, squat physical build of the Neanderthal.

But it was not only the characteristic weight/height body ratio but also the length of the rump in relation to the length of the limbs, seen in the relatively short upper thigh bones, which made them similar to Neanderthals. Were these people a type of Neanderthal? It is interesting to note that a comparison was made during the examination between these people and the primeval people of Siberia and Australia.

Were there Neanderthals on the islands of Southeast Asia? There is a large volume of literature documenting Old Stone Age artefacts on the northern part of Luzon (inter alia Königswald, 1956; Ronquillo, 1981). These finds were hotly debated because they conflicted with the prevailing time model for the settlement of western Pacific region and would mean that maritime travel already took place in the Old Stone Age. However, the waterways, where they existed at all, were far narrower than they are today, as the level of the sea was much lower than it is now.

As part of an archaeological study programme carried out by the Philippines University in central Luzon, numerous new finds of hand axes, produced using the Neanderthal Levallois technique, were documented (Mijares, 2001; cf. Pawlik, 2001). Thus in an area in which living "virtual Neanderthals" were found, there are also tools resembling Neanderthal artefacts from the Middle Stone Age. Many stone tools were found and these were assigned by European scientists as definitively belonging to the Acheulean period (*Homo erectus*), i.e. the older Stone Age (Pawlik, 2001).

The Levallois Technique ascribed to the Neanderthals is an exception in southeast Asia and otherwise has only been found in the Leang Burung 2 burial site in Sulawesi, Indonesia, where it was dated to only 19,000–31,000 years old (Glover, 1981), actually over 100,000 years too young, as this is in the age of early-modern Man.

In *Nature* (Vol. 77, 23 Apr 1908, p. 587) an unusual find was reported which had been described in detail with photographs by the *Gazette of the Polish Academy of Sciences in Cracow* (1908, pp. 103–126). In a grave in Nowosiolka, Poland, a skeleton had been found together with chainmail armour and several iron spearheads. In the clearly competently written article (Stolyhwo, 1908), this skeleton is compared with similar finds from the pre–Scythian Kurgan culture which lived in the Ukraine and in southern Russia up the Urals in the period from 500–300 B.C. and constructed buildings from early cyclopean stone. It is interesting and controversial to note the detailed description and measurement of the skull, which bears a great resemblance to that of a Neanderthal. An Old Stone Age skull in an ancient grave beside a coat of mail appears unthinkable and it is open to discussion as to whether this was a modern human whose skull possessed Stone Age proportions (Stolyhwo, 1908, pp. 103–126).

Fig. 26: Between Ourselves. Comparison of an acknowledged Neanderthal skull with the drawn shape of of a skull found in a grave beside chainmail (Stolyhwo, 1908, pp.103-126).

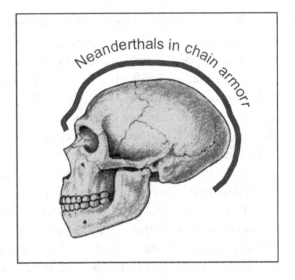

If we assume from the description of the skull in *Nature* that the Nowosiolka skull belongs to a Neanderthal, the counter-argument is that Neanderthals did not work iron and this cannot, therefore, have been a Stone Age man. Why are only tools ever found? If Neanderthals are tens of thousands of years old, then one can only find their stone tools because iron would have corroded long ago, would have dissolved into thin air, so to speak. In nature, iron artefacts survive only a few thousand years, if at all.

In my book *Mistake Earth Science*, I already discussed in detail the fact that far too few stone tools had been found from the Old Stone Age which supposedly continued for hundreds of thousands of years. If one shares out the stone tools found in France amongst 4,000 generations, one ends up with only 15 tools per generation for the whole of France. Precise examination of the French find site Combe Grenal (Dordogne) by the American archaeologists, Louis and Sally Binford (1966), showed that up to six different sets, each with eight individual implements – in other words almost 50 different individual implements – were used for various purposes (*American Anthropologist*, Vol. 68, 1966, p. 238ff.).

Working from this basis, one would find one specialised implement per generation at most in the whole of France.

Old Becomes Young

In *Mistake Earth Science* (2001, p. 197), I had already proposed that the period from *Homo erectus* to the present, via the Neanderthal, be reduced from 800,000 to a maximum 5,000 years and I concurred with the opinion of Gunnar Heinsohn (2003, p. 85 and 131).

Having assessed stylistic, ethnological and paleographic data for the emergence of modern Man, Heribert Illig (1988, p. 145ff.) suggested that this timespan be reduced still further to only 4,000 years and he drastically reduced the age of the late Aurignacian and Magdalenian periods placing them in the period 3000–2000 B.C. (ibid. p. 155).

Although two new finds of larger skull fragments should be mentioned (Berg, 1997;

Czarnetzki, 1998), the overall number of meaningful Neanderthal finds in Germany is still very sparse. In addition, a secure stratigraphic and archaeological underpinning for these finds is possible only in exceptional cases.

The two Neanderthals identified in 1967 from finds at the former Lahn Valley cave Wildscheuer (Knussmann, 1967) have held an assured place in the literature of human history since they were first publicised (inter alia Probst, 1999, pp. 356 and 376). The two skull fragments (Wildscheuer A and B) were said to be 75,000 and 60,000 years old.

In June 1999 the first comparative analysis was undertaken at the Wiesbaden Museum (Turner et al., 2000). The study showed that the two Neanderthal skull fragments from the Wildscheuer Cave were identical with the skull bones of a cave bear (*Ursus spelaeus*) found at the Scladina Grotto in Belgium: the finds which for 32 years had been celebrated as Neanderthal relics were, in fact, cave bears!

Sadly, this is not the only error. A few Stone Age human bones were found in the caves of the Swabian Alb. It is an area littered with Old Stone Age artefacts, such as the eight find sites from the middle and younger Old Stone Age in the famous Vogelherd Cave in Lonetal. The newer finds were figurative representations of peculiar perfection, which were supposedly carved 40,000–30,000 years ago from ivory and which were celebrated worldwide as Ice Age art.

It was not believed that these finely–wrought works of art of the Aurignacian culture could have been made by Neanderthals – and this disbelief is not completely without foundation. After all, in some correspondingly old layers, the bones of anatomically modern humans were also found.

The skull which was found by Gustav Riek in 1931 in the Vogelherd Cave, in one of the most prolific Aurignacian layers, peppered with "Ice Age art", is also world-famous. He himself said that the find layer was "completely undisturbed". For more than 70 years, this assertion was not questioned and was accepted as a matter of fact in all publications.

On 8 July 2004, a report appeared in *Nature* (Vol. 430, pp. 198–201) in which Nicholas Conard, a successor of Gustav Riek to the Chair of Prehistory and Early History at the *University of Tübingen*, presented the radiocarbon dating results for the Vogelherd skeletal remains. A scientific sensation: the six bone fragments examined are not around 32,000 years old but only 5,000–3,900 (ibid p. 198). The Old Stone Age Aurignacian skull was transformed into a modern human from close to the end of the younger Stone Age.

Since Gustav Riek had explicitly confirmed that the layer in which the skull had been found was undisturbed, the corresponding layers must now also be reduced in age: the middle and younger Stone Age layers are now Young Stone Age, with their ages reduced by over 27,000 years. As such, the alleged "Ice Age" art embedded with the skull in similar layers is also correspondingly younger, i.e. from the younger Stone Age.

But no: the world famous sculptures, around twenty in all, are to keep their age, because they are the oldest known authentic man–made works of art. To preserve this position, it is now said that contrary to his unambiguous confirmation, Riek must have been confused: in Conard's opinion, Riek supposedly made a mistake and the bone fragments must therefore have been buried in the far older Aurignacian layers during a Young Stone Age burial. Since these geological layers have been dug away, there is naturally no proof for the claim that Riek was wrong. On the basis of faith alone, the alleged Ice Age sculptures are to retain their documented age, as otherwise the history of Man would have to be rewritten. This is a trick of legerdemain masquerading as scientific method!

The deposits in the Vogelherd Cave are around 2.4 metres thick, whereby the deepest layer is said to be over 350,000 old. The boundary between the older Aurignacian (Layer V) of only a good 30,000 years (according to Zoth, 1951, p. 264) lies at a depth of about 1.5 metres. If, on the basis of the skull find, one reduces the age of this layer from 32,000 to 5,000–3,900 years, the geological layers above it also become younger.

In fact, the upper layer (Layer I) is already supposed to be 4,500 years old and date back to the younger Stone Age. Accordingly, Layers II to IV must also be much younger. And is it correct that remaining bottom 90 centimetres (beneath the rejuvenated Layer V) can really be far more than 300,000 years old? Is it the case that all the layers are, in fact, 5,000 years old rather than 350,000 and were deposited rapidly by floodwaters?

In any event, a key find of the first modern human in Europe with whom the sudden emergence of cultural modernity through the appearance of Cro-Magnon Man is supposed to be proven, is now pulled into historically tangible reach as if by a time machine. Did the first modern human appear only 4,000 years ago? The Vogelherd find is, in fact, the last in a series of key discoveries all of which have been truly overturned in recent years through modern dating.

The question of who created the earliest art is once again thrown open completely. "Further candidates were lost only recently", says Thorsten Uthmeier of the *University of Cologne*. "For example the skeleton from the Cro-Magnon cave in the Dordogne has been re-dated. Like the Vogelherd bones, it was taken as proof that the art of the early Aurignaciens was created by modern humans. But being only around 25,000 years old, these remains are too young. Cro-Magnon Man could not have made a name for himself in the art world before the (later) Gravettian phase" (*Frankfurter Allgemeine Zeitung*, 11 July 2004, p. 51).

The oldest material from modern humans in Europe is the Mlade Skull from the Czech Republic which is currently said to be 32,000 years old. In other words: this is the first time that the modern human becomes tangible in Europe. These bones have not, however, been dated using new methods …

I suggest that with the rejuvenation of the bones, the age of the Aurignacian works of art also be reduced on simple grounds of logic, from over 30,000 years to 4,000 or, at most, 5,000 years, since they are in similar geological layers to the Vogelherd bones.

By doing this, however, all of the epochs of the geological layers which lie above the find layers can be compressed into a few manageable millennia.

In other words: the existence of early–modern Man from the Aurignacian culture should be reduced to 4,000 years, since the Middle Stone Age is a phantom period, as will be demonstrated later on. As such, the Old Stone Age was a short, traumatic cultural phase after the Deluge, around 5,000 (or perhaps only 4,500) years ago, if we take my contentions in *Darwin's Mistake* as our basis.

If scientists quarrel and now want to ascribe the Aurignacian art works to the Neanderthals, this can also lead to a dead end, because now even the few firm proofs of a certain cultural modernity amongst Neanderthals in the Old Stone Age have to be made younger and thus irrelevant. Neanderthal bones, which were found in France together with the sophisticated implement culture of the so-called Châtelperronian period, are threatened with a similar fate to the Vogelherd bones.

After new examinations of the old finds, Jean-Guillaume Bordes of the *University of Bordeaux* drew attention to this point in Blaubeuren. According to him, there are serious doubts are to whether the Châtelperronian implements really have anything to do with Neanderthals.

"When it is a question of human, Neanderthal and the origins of art, it appears that we know less and less the more closely we look. But it is also a question of the subject matter which is almost without rival in encouraging the imagination to fill the immense gaps in the available data" (*Frankfurter Allgemeine Zeitung*, 11 July 2004, p. 51).

The eight find layers with Aurignacian works of art from the Vogelherd Cave and neighbouring caves should simply be made younger, just like the bone fragments. This way, it remains the art of modern humans. The myth of supposed "Ice Age" art will have to be given up, especially since there was no "Great Ice Age" (discussed in detail in *Mistake Earth Science*).

This jumble of dates starts to make sense when one pushes the Neanderthal into the same timeframe as Young Stone Age Man, i.e. 5,000 years ago at most. Can't be done that easily? If not only the age datings of early-modern Aurignacian bones but also those of Neanderthals were faked, then it's relatively quick and easy.

Crime Scene: Frankfurt University

In *Darwin's Mistake* the Stone Age with the Neanderthal is represented as a purely post-Deluge, wrongly interpreted epoch. According to the additional arguments presented in *Mistake Earth Science* the Middle Stone Age is a phantom period and the Old Stone Age is only a short phase of 5,000 years at most (Zillmer, 2001, p. 197ff.). To the horror of the paleoanthropologists, this opinion has been lent support by a report which appeared in August 2004:

"Numerous Stone Age skulls in Germany are probably far younger than previously claimed. The Frankfurt anthropologist, Prof. Reiner Protsch von Zieten, had overestimated the ages of important finds by tens of thousands of years" reported the news magazine *Der Spiegel* in connection with new radiological datings carried out by *Oxford University.*

Analysis using the so–called radiocarbon method (C–14 method) revealed that instead of 30,000 years, some of the skulls were only a few hundred years old. "Anthropology must now draw a new picture of the anatomically modern Man in the period from 10,000 to 40,000 years ago" said the Greifswald-based archaeologist, Thomas Terberger. The former head of the *Hamburg Helms Museum*, Ralf Busch, confirmed that the Neanderthals from Hahnöfersand are only 7,500 rather than 36,300 years old. The Binshof–Speyer Woman is not 21,300 years old but lived around 1300 B.C. The skull found at Paderborn-Sande ("the oldest Westphalian") is not 27,400 years old but belonged to a person who died around 1750 A.D. Sadly, after "weeding out the bad apples" there are barely any meaningful human finds from the period between 40,000–30,000 years ago, said Terberger. "The oldest bone find in Germany is ... now a skeleton from the central Klausen Cave in Bavaria, which is 18,590 years old" (*dpa*, 16 Aug 2004, 17:59h).

Prof. Reiner Protsch is said to be responsible for the incorrect dating. A further key find for Stone Age culture, the "Kelsterbach Lady" who, said by Protsch to be 32,000 years old, was considered to be the oldest known, anatomically modern human after the Neanderthals in Europe. Thus the theory that only our forerunners, the Neanderthals, lived at this point in time in Germany, is overturned. Now the key find has disappeared and so cannot be dated anew. Investigations were launched by the criminal police and the public prosecutor. Without this skull the co–existence of Neanderthals and modern humans in Germany cannot be confirmed.

We already met Protsch in connection with the false dating of the Reck bones from the Olduvai Gorge. Former staff reported to *Der Spiegel* news magazine that the scientist "simply dreamt up" dates. Conclusion: Everybody knew that false ages had been assigned to the Stone Age skulls. Since the imaginary ages fitted in perfectly with official theory, the scientists believed that the end justified the means in the interests of science!

It is astonishing, or perhaps significant, that the extremely young age datings which, in the worst case may be as far apart as nearly 29,000 years, failed to create a stir amongst experts or in the universities. Only after Der Spiegel picked up on the subject did the directors of the *University of Frankfurt* immediately create a commission for dealing with scientific misconduct.

However, if at all, only Prof. Protsch will be called to account. The parts played by his helpers, co–authors and staff are minimised. In his justification, Protsch claimed that the possible wrong age datings may have been caused by contamination of the finds, perhaps by microorganisms. A seven year old bone smeared with mineral oil can easily be dated to thousands of years old. As an anthropologist one must know how easily a find can be made older so that it fits into the right period.

For Protsch, age datings are supposedly experiments, not definitive statements and therefore not fakes. Does this mean that anthropologists have jester's licence? Yes it does, because his (mental) experiments are pure invention. The radiocarbon dating equipment had "never been used" before 1981 (*Der Spiegel*, 34/2004) and the laboratory had no calibration parameters: the supreme expert on Stone Age skulls was incapable of undertaking a professional radiocarbon dating, but continued to write human history with his fantasy age datings for decades, like a *Grimm's fairy story*, which was then imaginatively presented by the mass media to an astounded public as "proven" fact.

But now that numerous Stone Age skulls have been pushed into a younger historical timeframe, let's take another look at the finds of sophisticated implements in the Tertiary Red Crag formations in England, (too old for human history) at a time when Lucy "was trying to toddle" (Lohest, et al., 1923, p. 67), which are controversial but are still recognised as authentic by a scientific panel.

In the textbook *Der Mensch im Eiszeitalter* (Man in the Ice Age) Josef Bayer (1927, p. 205) suggests that the Red Crag, if not also the Coralline Crag, be included in the Quaternary (Diluvium), i.e. the age should be considerably reduced. Thus the (anomalously old) stone implements would at least fall into Europe's Homo age. By telescoping time, the finds would be less controversial in terms of human history. But then the age of both the layers in which these implements were found and the layers above must then also be reduced – because the two timescales of geology and evolution are indissolubly linked, come hell or high water. Josef Bayer "proved that the prevailing conventional wisdom was suffering from a multiplication of the actual time periods, so that the absolute length of time of the Quaternary (Diluvium – HJZ) is also much shorter than previously assumed…" (Bayer, 1927, p. 452).

Looked at from this angle, individual cases in which finds have been considerably reduced in age become understandable. Herman Müller-Karpe went so far as to reduce the age of a Inuit bone scratcher in the north Canadian Old Crow region by a factor of 20. Using the AMS method (an improved radiocarbon method) it was shown that the bone was not 27,000 years old, as a 1960s dating had established, but that it came from an animal which died only 1,350 years ago (Strauss, 1991, A12; cf. Heinsohn, 2003, p. 83). Must other finds with earlier age datings be re–examined, and must all of them have their ages reduced many times over? But even with the AMS method, the dates are still too high, as is suggested by the dating of cave drawings.

The rock drawings in the Chauvet Grotto (Ardèche Valley, France) were first classified by Jean Clottes as belonging to the middle of the young Paleolithicum (middle of the younger Old Stone Age) on the basis of stylistic features. "When the AMS data (from pigment particles and organic material from rock crusts – HJZ) showed that the paintings dated to the Aurignacian period, the Chauvet Grotto was seen as proof that there had already been outstanding works of art and artists in France in the early young Paleolithicum – in short, art history had to be rewritten and the classic methods of rock drawing research had failed.

Placing blind trust in the 'exact' sciences goes so far that even an obvious 'key fossil' from the end of the Solutrean and beginning of the Magdalenian, bearing a newly discovered "claviform marking" was declared to be a simple symbol without meaning and was thus pushed aside even though it would fit in exactly with the expected archaeological timeframe" as Dr. Christian Züchner (2000) of the Institute for Prehistory and Early History at the *University of Erlangen-Nuremberg* gives us to consider.

In other words, the age datings using the AMS method produced an age twice as high as that given by archaeological dating. Are the new datings too old, especially since the rock drawings dated to the Aurignacian culture period with the same details reappear during the Magdalenian after an absence of 15,000 years (Züchner in Quartär, 51/52, 2001, pp. 107–114)?

When one considers the freshness of these fantastic cave paintings, it appears that the archaeological datings are also too high, in particular if one bears in mind that during the last Ice Age, drawings depict heat–loving animals and naked people. Do these pictures belong to this newly crystallising period 5,000–4,000 years ago?

The phase of stylistic development of certain rock painting motifs also appears to be far too long. Gunnar Heinsohn (2003, p. 87) rightly asks (cf. photo 52): "Whether the modern human is not being overestimated if he is given not 20,000 years but only around 1,000 for his artistic development as far as the horse's head of the 'classic' period of the late Magdalenian"? 9,000 years to achieve the refined representation of a male bison? "Couldn't that have been achieved in 900 or 700 years or possibly even less time?" (ibid, p. 87).

In *Science* (Vol. 283, 26 Mar 1999, pp. 2029–2032) the question is raised as to whether a lack of finds might be responsible for false interpretations. If one interprets the presence of Stone Age implements not by the standard model but using a demographic factor, a telescoped time model is the result.

As a result, the implements would have been produced more than 15,000 years later than the standard model suggests. Consequently, the implement finds do not signify the sudden appearance of groups of modern humans and thus a displacement of the Neanderthals, but are much more likely to document gradual progress in the Old Stone Age, which was sustained by groups of people around the Mediterranean, particularly in southern France and the north coast of Spain.

This study represents an alternative interpretation of archaeological finds to the conventional timeframe. Since a date cannot be put on when the implements were made, the implements could also be made younger, along with their newly–dated owners, to a period around 5,000 years ago. Thus the stone implements bear witness to a continuous development of peoples without the displacement of the Neanderthals only 20,000 years later during the Young Stone Age. But aren't the stone implements in geological layers whose age is known? Let's take a look at cave stratigraphy.

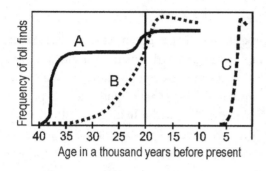

Fig. 27: Matter of Opinion. *Current opinion holds that advanced stone tools suddenly appeared when early-modern peoples arrived in Europe (A). If one considers a demographic model (B), these same finds indicate a slow (gradual) development (Science, Vol. 283, 26 Mar 1999, p. 2029). Using model B, the age of the geological layers would also have to be "stretched" (time inflation). Without this stretching of time and analogous to the skulls of early modern humans whose age was reduced by up to 28,000 years, Zillmer claims that these stone tools developed explosively (as in model A), but only around 5,000 years after a natural catastrophe.*

Cut Sharply

The founder of Old Stone Age chronology, Gabriel de Mortillet (1821–1898) was of the opinion that the Young Stone Age followed directly on the heels of the Old Stone Age. Edouard Cartailhac (1845–1921) was of the view that large parts of the European continent were uninhabited for a period between the Old and the New Stone Age.

The layers from this period are very meagre and contain almost no relics, which is why they do not suffice for a Middle Stone Age of 4,000–3,000 years. It is not even clearly proven that the settlements which are said to belong to the late phase of this period were occupied all year round (Champion et al., 1984, p. 103).

There are without doubt layers above the Magdalenian which conceal new finds, but which do not represent a fully developed Young Stone Age. This phase would, in stratigraphic terms, be an extremely short phase of human history: Long vertical stratigraphies are rare everywhere (Champion et al., 1984, p. 97) and in England there is no record of a single burial for the good 4,000 Middle Stone Age years (ibid., p. 108). Doubts about the existence of a Middle Stone Age lasting several millennia, which were already expressed by Heribert Illig (1988, p. 29, 160) are confirmed by stratigraphic analysis.

The embarrassment of being unable to archaeologically confirm a Middle Stone Age of several thousand years even leads archaeologists to undertake attempts at falsification and concealment. "Thus an excavation report without a Middle Stone Age was originally presented with one of the best German stratigraphies – the Ilsen Cave in Burgberg, near the Thuringian town of Ranis" (Hülle, 1939, p. 105ff.). In 1961 a representation was produced showing a paper–thin Middle Stone Age layer which however was directly below the Bronze period, with no sterile intermediate layer, so that it could still be classified as either young or Old Stone Age and could not represent an epoch in itself.

The original excavator of the Ilsen Cave "would not accept the manipulation of his life's work" and presented a stratigraphic sketch without a Middle Stone Age (Heinsohn, 2003, p. 106).

There is also a layer missing from the aforementioned Vogelherd Cave which could be ascribed to the Middle Stone Age. In the internet lexicon *www.akademie.de* (as at 20 Dec 2004), for example, it is floridly stated: "Use in the Young Stone Age was preceded by an inspection in the Middle Stone Age". This assumption is pure invention but at least the Middle Stone Age has been mentioned for the benefit of uncritical readers.

Do geological layers – the stratigraphy – bear witness to a gradual transition from one cultural stage to the next or are the layers clearly separated as if cut with a knife? Gunnar Heinsohn has dedicated considerable effort to this subject and points out that "the stratigraphs of many caves... (contain) so-called sinter layers (chalk tufa) and loess layers (also known as clay layers) between the cultural layers. These intermediate layers are largely sterile in archaeological terms and have led to debates which have not yet been resolved up to today" (Heinsohn, 2003, p. 74). The sterile intermediate layers are a sign of short–term water penetration.

A very good example is the profile of the Abri de Laussel Cave (Bayer, 1927, p. 57). Layers lying apart from each other suggest that there was a direct, abrupt transition from *Homo erectus* (Acheulean) to Neanderthal (Mousterian), while all other cultural layers up to the Solutrean, with their sterile intermediate layers, indicate clear separate periods. The Magdalenian, the middle and the Young Stone Age are entirely absent. The profile of the Trilobite Cave also shows sharply separated cultural layers from the Mousterian to the Young Stone Age, but is without Acheulean (*Homo erectus*) and, again, without a Middle Stone Age. It appears that this cave records the abrupt transition from Neanderthal–Mousterian to modern Aurignacian Man (Cro-Magnon).

What consequences does the sharp separation of cultural layers have for the development of humans? The theory of multi–regional emergence of modern Man underlines the gradual transition from Neanderthal to early–modern Man in Eurasia. This is why the multi–regionalists draw attention to the fact that there is little anatomical difference between the two, although the Neanderthal is indeed very different from modern Man in view of his squat stature. However, since the European stratigraphies described are mainly sharply separated from each other, a gradual (according to Charles Darwin) evolution would appear to be out of the question. Stratigraphic findings seem to favour the displacement theory. On the one hand the frequently observed sharp separation of these cultural layers, proves that there cannot have been a gradual transition from *Homo erectus* to Neanderthal when one considers the enormous anatomical leap in brain size from 1000 to more than 1500 cubic centimetres with the accompanying change in the shape of the skull. On the other hand, with the transition from *Homo erectus* (Acheulean) to Neanderthal (Mousterian) and from Neanderthal to modern Man, one is sometimes surprised by the simultaneous appearance of relics of both cultures.

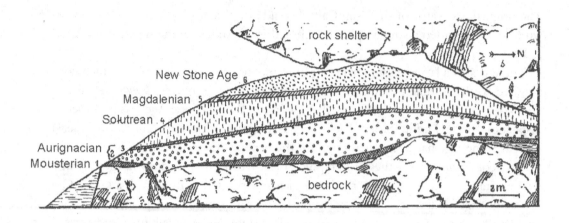

Fig. 28: Loess Layers. *The stratigraphy of the Trilobite Cave near Arcy-sur-Cure shows direct transitions between Neanderthal Mousterians and early-modern Aurignaciens. Middle Stone Age layers are absent. The "cultural layers" are separated by sterile intermediate layers. The loess layers (1,2,4 and 6) were washed into the cave and each in turn covered by clay and sinter layers (tufa) (3,5) formed by water seeping out of the fissures in the rock roof mixing with the dissolved lime contained in the water. Not 30,000 years but 100, and at most 1,000 years would be sufficient for this formation. Layer composition assumed from Bayer, 1927, p.58, cf. Heinsohn, 2003, p.67ff. Similar stratigraphic layers too those of the cave can be found at the foot of cliffs, in the form of tiered terraces*

Why should a sudden anthropological transformation have taken place in any case, when for almost 100,000 years from early to late Neanderthal, "there was no indication of any evolution into modern Man" to be seen (Gambier, 1989, p. 207)? The technical standard of implement manufacture also stagnated during this period only to be suddenly revolutionised. Since in archaeological terms the transition from Neanderthal to modern Man is stratigraphically direct, the rapid displacement of the Neanderthal appears to best fit in with the finds and the facts. In his supposed homeland too, be it in Africa or in Israel, no fine transitional stages from Neanderthal to modern Man have been proven. Within the framework of prevailing teachings, the debate between the multi–regionalists and the advocates of displacement theory rages on.

For this reason Gunnar Heinsohn posited the following: Neanderthal women have babies who were immediately fully–developed modern humans (Heinsohn, 2003, p. 61). This would explain why in one and the same layer, implements of supposed parents (Neanderthals) and their children (modern humans) are found together whilst in the layers above there are only relics of modern humans. The offspring did not kill off their parents but lived with them until they died a natural death (Heinsohn, 2003, p. 126).

Heinsohn's explanation for this sudden "hypo–macroevolution" – which took place in huge leaps – is an electromagnetic mutation catastrophe at the end of the Neanderthal era. However, such a catastrophe would also have meant the sudden transition of *Homo erectus* into Neanderthal at the same time. There have indeed been major cataclysms in the past accompanied by electrical discharges and increased radioactivity – something which has already been mentioned by Immanuel Velikovsky (1980, pp. 258–260) in connection with the effects of the Hiroshima bomb.

Nevertheless, despite intensive laboratory tests, it has never been proven that increased radioactivity can lead to advanced development of species. Rather it is responsible for the extinction and deformation of species so that, for example, fruit flies exposed to radiation suddenly have more than their usual four wings. At the end of the Ice Age (= end of the Deluge) around 80% of all animals species, such as mammoth and giant deer, are said to have disappeared because of climate change or human depredations.

However, on the basis of fossil finds in Siberia, the time at which the giant deer became extinct has been reduced from 10,500 years ago to 7,700 (*Nature*, Vol. 431, 7 Oct 2004, pp. 684–689). Mammoths too survived. A dwarf mammoth, measuring 1.8 metres at the shoulder, was discovered on Wrangel Island, 120 miles of the northeast coast of Siberia, inside the midsummer pack ice boundary. These dwarf mammoths supposedly existed up to 3,700 years ago (Lister, 1997, 34f., and *Nature*, Vol. 382, 1993, pp. 337–340). How did mammoths get there in the first place? Why were mammoths living on what are now hostile islands? Were they taken by surprise by the rising of the sea level with accompanying flooding of the flat continental shelf in the North Polar Sea and end up being trapped on a mountain top which later became an island?

In America a complete steppe megafauna ranging from horse and camel, to giant sloth to mastodon was wiped out by natural catastrophes and abrupt climate changes. Are the age datings involved here correct? The news agency CBC broadcast the report that, contrary to accepted teachings, North American Indians were familiar with horses, which they hunted and slaughtered (*BdW*, 11 May 2001).

But let's stay with human development. On the basis of stratigraphic findings, a sudden mutation within a single generation (Heinsohn, 2003, p. 74) appears to provide a way out of the dead end in terms of the stratigraphic, archaeological, cultural and artistic conditions discussed – but the question remains as to whether there is not some other explanation for the sudden emergence of *Homo erectus*, Neanderthals and modern humans because for me, the idea that a modern human baby could spring fully formed from its Neanderthal mother's womb seems quite incredible.

One should also remember that the modern human is said to have left Africa 140,000 years ago, to have arrived in Israel 100,000 years ago, where he took over from the Neanderthal. However, in the caves of Amud and Kebara in Israel, the remains of Neanderthals which were only 60,000 years old have been found. Modern humans from this period have not been found. Did the Neanderthal displace the modern human again?

It seems strange that at this time the Neanderthal used the same methods for producing implements, used fire and practiced burial, just like the modern human who lived in the same caves before him in this area (Arsuaga, 2003, p. 301f.). Or did modern humans leave the area for quite different reasons, perhaps because of some drastic change in climatic conditions?

Nomadic Carnivores

The analysis of Neanderthal bones by an international team of researchers showed that the Neanderthal's diet was almost exclusively of meat and that is why they were such successful hunters. If they had fed primarily from dead meat they would have had to have eaten vegetable food as well, in order to survive (Richards et al., in *PNAS*, 20 June 2000, Vol. 97, pp. 7663–7666). A further study of bone relics of early modern humans found in the Czech Republic, Great Britain and Russia compared them with those of Neanderthals who were roughly contemporaneous. Whilst it appears that Neanderthals were exclusively big game hunters, researchers believe that early modern humans could not only fish, but also preserved fish by drying. In addition, early humans also caught birds using traps.

The Neanderthals spent most of their time hunting. When their prey changed its territory or a new hunting competitor appeared, the Neanderthals did not know where they should next try their hunting luck (*PNAS*, 22 May 2001, pp. 6528–6532). For the archaeologist Curtis Runnels, the decline of the Neanderthal was sealed with the disappearance of the great herds of bison and giant deer, caused by climatic changes (*Science*, Vol. 303, 4 Feb 2004, p. 759).

If the Neanderthal was really a carnivore and followed the game, a nomadic lifestyle may be inferred. This differs from the sedentary early humans living at the same time. Right up to the present day there are different ethnic groups living alongside each other, on the one side settled, on the other nomadic. Interbreeding does not normally take place, except when the nomads are forced to become settled because of the drawing of artificial borders. But sexual contact and isolated instances of hybrid births are nevertheless both possible and even probable. The supposed extinction of the Neanderthals, which is often only supported by the absence of "typical" Neanderthal implements in certain geological layers, could simply indicate that these nomadic people followed the animals as a result of climatic changes and sought new places to live.

In the Zagros Mountains in the southwest of Iran, Neanderthals hunted wild sheep and goats. Before these wild animals became extinct (or migrated because of climate change?), says one study, modern Man domesticated them (*Science*, Vol. 287, 24 Mar 2000, pp. 2174–2175). Might it not also be possible that the Neanderthal did not exclusively hunt sheep and goats but that, as nomads, they herded the animals and went hunting in between times? Sedentary farmers would then have arrived with domesticated animals on foot of climate changes? It is interesting that early domestication in the form of herding supposedly began only 10,000 years ago and actual breeding did not start until 1,000 years later (Mareau, 2000, p. 2174). This transition is said to have played a role in the extinction of the Neanderthals and the emergence of modern Man. It is not always necessary to apply evolution theory's "Survival of the Strongest (Fittest) Law". During the so–called Stone Age population density was so low that there was sufficient space for all people and large living space for various life forms. Certainly, in some places there would have been a continuous development from a nomadic to a sedentary way of life, caused by the drastic worsening of the climate and the diminution of the steppes, for example after the Deluge. This transition would have been accompanied by a natural hoarding of existing meat resources and thus domestication of animals.

Neanderthals did not live in caves although, as may be seen by the example of the Kartstein Cave near Euskirchen in Germany, a hut was built into a recess in the cave wall. The inhabitants would have had their hearth, whose ash remains were found, in or in front of the hut. In the open air, the Neanderthals built huts or tents out of wood, large bones and pelts.

Sensational traces of Neanderthal huts made of animal skins on a frame of branches and mammoth bones have been found in Russia and the Ukraine. Outside the hut was decorated and weighted down with further massive mammoth bones and tusks. This area was a fertile savannah landscape, also home to the the pre–Scythian, grave mound building Kurgan culture (Kurgan is Turko-Russian for grave mound) in 5000–3000 B.C. This is the period into which the bones of early–modern Man of the Aurignacian culture have been placed with new datings. I would like to remind you of the grave of a Neanderthal–like skeleton, found in Nowosiolka, Poland beside chainmail armour and several iron spearheads. The comparison of this skeleton with those of the Kurgan culture (Stolyhwo, 1908) is interesting for geographic reasons alone because the Neanderthals coming from the freezing North and early–modern Man (Aurignacian) were displaced from the homeland areas of the Kurgan and/or Scythian peoples north of the Black Sea, before the latter existed. Huts have been discovered in the Russian tundra, in Pushkari for example, whose architecture resembles those of the Neanderthals but which have been classified as having been built by modern humans in the younger Stone Age.

The Saiga antelope too, which lived throughout the tree savannah which stretched from southern France to the Bering Sea during the last (alleged) cold period, today now lives on in the same fallback area as the former "Stone Age Man" in the region of the Black and Caspian Seas, whilst the mammoths, elephants, hyenas and rhinoceroses who also lived in Eurasia became extinct. There were also lions during Europe's alleged Ice Age, as illustrated by the supposedly 32,000 year old drawings by early-modern humans in the Cauvet Cave in France.

Fig. 29: Huts. *The sensational discovery of Neanderthal huts was made in the Ukraine (from: Dorling, 1994, p. 18). Similar huts are built by modern man, such as those of the Pushkari in the Russian tundra.*

When comparisons are drawn between the Neanderthals and the people of the steppes in the Black and Caspian Sea regions, one must remember that the face and body of the Mongol are built in such a way that they offer protection against extreme cold. The body and the head in particular are as rounded as much as possible. The body's surface ratio is small compared with its volume, to reduce heat loss. The nose and the nostrils are small to avoid the risk of freezing. The eyes are protected by lids which form veritable sacs of fat and they leave only a small opening which allows good vision but protects against the icy wind of the Siberian winter (Cavalli-Sforza, 1999, p. 23). Slit eyes appear to be an adaptation to climatic conditions.

If, unlike Cro-Magnon Man, the Neanderthal was supposed to have been better adapted to the colder climate, this means that the Neanderthal either could not leave the cave or "permanently wore clothing to protect himself from the cold. To produce these clothes, animal skins were spread out and freed of fat and sinews using hand axes. After drying, the skins were then sewn to produce the given article of clothing (Dorling, 1994, p. 18).

The discovery of steel sewing needles revealed an interesting fact. This fossil artefact was dated to an age of 26,000 years (Johanson, 1996, p. 99). Even if I reduce the age in keeping with the Neanderthal skull to, at most, 5,000 years during the Young Stone Age, this remains a sensational find because according to prevailing teachings, Europe's Iron Age started only around 2,700 years ago in the Hallstatt period. This find underlines the opinion I expressed in my first books, that the breakdown into Stone, Bronze and Iron Ages is essentially wrong.

The lifestyle of the Neanderthals resembled that of the Inuit (or, as they call themselves, "Yuit"), who are known more widely by the abusive Indian name of "Eskimo" which translates more or less as "raw meat eater". Their habitat is the Arctic, Siberia, Greenland and northern Canada. Although the distances between these areas in some instances is more than 5,000 km as the crow flies, the Inuit everywhere have a similar language and culture. The only exception are the Inuit of southern Alaska and the Aleutians, who have adopted cultural elements from the Indians of the northwest coast.

In Arctic regions one more often sees people with compact body structure or a generally large body volume in relation to body surface. These characteristics are very useful for the better conservation of body heat. Lighter skin is probably also less sensitive to frost than dark skin. All of these characteristics may be observed most frequently amongst northern peoples such as the Inuit, the Sames (Lapps) and the Siberians. Nevertheless, all of these human groups demonstrate the same physical reflexes, for example, the trembling of muscles to create warmth. Man's main adaptations to climatic conditions are both biological and of a specific cultural nature, through clothing, housing and the use of fire.

The skeletal remains of Neanderthals demonstrate a particular combination of characteristics. Alongside skull and lower jaw, these relate to the body structure in general. Neanderthals were about 1.6 – 1.7 metres high and of a powerful, squat stature. The men weighed about 70 kg, the women around 55 kg. The heavy skeleton corresponds with a strong musculature. In proportion to the torso the limbs were quite short, similar to today's Arctic peoples, the Inuit and Lapps.

These people were obviously well–adapted to a cold climate: "In today's humans the body proportions are closely linked to the geographic latitude of their habitat". To establish this, the weight/height ratio and the proportions of the limbs to the torso are measured. Using these measurements, average values for different ethnic groups are obtained, so-called main components. As shown in Fig. 30, people from northern countries are on the far right of the diagram with regard to the first main components: The body structure is more compact, the colder the climate. The Neanderthal was obviously well–adapted to harsh weather conditions (Hublin, 2004, p. 58).

Is a mistake being made in assuming that the above–mentioned similarities must mean that Neanderthals and Cro–Magnon Man shared a common ancestor? It has already been shown that implements of Neanderthal and early–modern Man can be found together or are discovered in a seemingly incorrect, reversed layer sequence – in other words contrary to the supposed direction of development.

The riddle of the Neanderthals seems to be solved if we stop thinking of the Neanderthal as a primitive forerunner or as a dead branch of the human evolutionary tree. In keeping with many publications by the expert, Erik Trinkaus (*PNAS*, Vol. 94, 1997, pp. 13367–13373 and *Curr. Anthropology*, Vol. 41, 2000, pp. 569–607) the differences between Neanderthals and modern humans are rather the result of adaption of physical characteristics to a cooler climate (*PNAS*, Vol. 100, 10 June 2003, p. 6926).

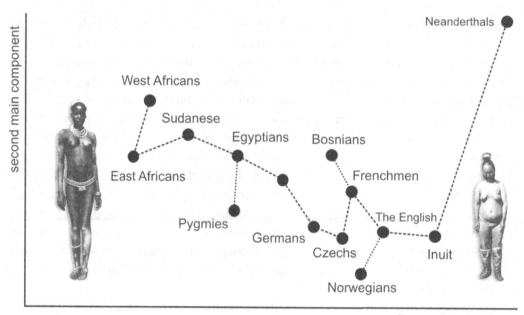

Fig. 30: Body Proportions. *The compactly-built Neanderthal was even better adapted to the cold than the Inuit (Hublin, 2004, p.58).*

Timothy D. Weaver (2003) examined the form and properties of the upper thigh bones of various peoples, both living and extinct. He found that people living today generally differ based on the climatic conditions under which they live: for example, East and South Africans and Australians in warmer, British, Aleut and Inuit in cooler climates. As such, the modern human falls into the first category and the European Neanderthal (Neanderthal 1 and Spy 2) into the second. This study confirms that the Neanderthal quite simply belongs to the (modern) people who are better adapted to colder/Arctic conditions. A special study of human body measurements even came to the conclusion that the body of the Neanderthal showed signs of hyperpolar adaptation (*Journal of Human Evolution*, Vol. 32, 1997, pp. 423–447).

The physical characteristics of the people who inhabit colder climate zones allow them to manage their body heat better than people from the tropics. According to a study published in *Science* (Vol. 303, p. 323) the difference is due to changes in mitochondrial DNA. Given the drastic reduction in age of the few Neanderthal skulls still regarded as "authentic", the Neanderthal appears to be not a forerunner of modern man or a separate species, but a type, simply a variation of modern Man who was particularly well–adapted to the cold.

This assumption is supported by the mentioned study: form and properties of the

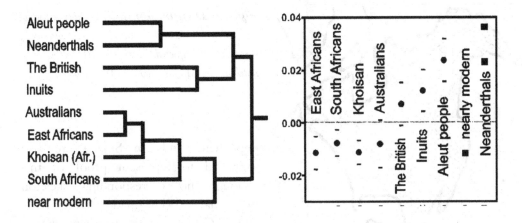

Fig. 31: Climate Adaptation. A comparison of the various functions (form) of an upper thigh bone showed that Neanderthal Man – similar to the Aleuts and Inuit - belonged to those peoples adapted to cold climates, whereas modern Man is adapted to a warmer climate. The dotted horizontal line shows the average values of today's peoples (Weavewr in: "PNAS", Vol. 100, 10 June 2003, p.6926).

upper thigh bones are not extreme in Neanderthals but resemble those of the indigenous peoples living today in the Aleutians.

Where do the Aleut come from? It is assumed that around 10,000 years ago, in other words after the Ice Age (= deluge and Snow Time after HJZ), Siberian nomads known as the Unangan (Aleuts) settled the Aleutian island chain which extended in a curve for around 2,000 km from the Alaska Peninsula to Siberia, separating the Bering Sea from the Pacific. The harsh weather conditions on the coasts of the Bering Sea forced the inhabitants of the region to become highly adaptable. The Aleut differ from the Inuit in that they have round faces.

These "Arctic Indians" belong to a language family known as the Eskimo–Aleutian. In western and northern Canada and as far as Greenland, Na-Dené represents a further language family. Both "belong to the language families of the Old World, the Eurasian and Dené-Caucasian respectively" (Greenberg/Ruhlen, 2004, p. 59).

Since as the eastern branch of Dené-Caucasian, Na-Dené is clearly different to Eurasian, Na-Dené cannot have split from Eskimo-Aleutian on the American continent but must have found its way there during a later wave of migration (Greenberg/Ruhlen, 2004, p. 63). After a first wave of migration from Siberia 15,000–12,000 years ago, a second is supposed to have taken place 5,000–2,000 years ago (*PNAS*, Vol. 98, 14 Aug 2001, p. 10021).

There is still disagreement as to where the first settlers in eastern Siberia came from. After a study comparing the skulls of various peoples, it was thought that an eastward expansion from the Old World via Siberia had taken place 200,000 years ago, and

*Fig. 32: **Distribution.** Prehistorically waves of emigration in North America to Cottevieille, Giraudet. Chequered: Eskimos (Hyperborians); hatched: Cro-Magnon; dotted: Asian peoples.*

implement finds in Siberia are also supposed to represent proof of this although no corresponding skeleton finds have been made (Derev'anko, 1998). Other experts proclaim that a migration wave came from the south, through Asia, possibly across India, Korea and China through to eastern Siberia and from there, turned around, so to speak, to move in a westerly direction across Siberia to Europe and in an easterly direction through Alaska to North and South America (Cavalli-Sforza, 1996, p. 109).

Let's take a short break and consider new thoughts and research findings. Within the scientific community too, some researchers are making an about turn with regard to the Neanderthals. It appears that there was no evolution in the sense of Darwin's theory of evolution, but that the Neanderthal, until recently regarded as our primitive forerunner is actually a modern human, whose body was quite simply adapted to Arctic conditions. This is an adaptation to climatic conditions for which we don't need any "fantastical" explanation or sudden mutational leaps. The macroevolution from primitive to modern Man, propagated as the climbing of rungs on an evolutionary ladder, is seen to be an error, carefully maintained for 150 years by science and the media.

There is another question: did America's early settlers come not from Siberia via the Bering Straits but from Europe via Iceland and Greenland? This connection via the Greenland Bridge, which I discussed in my book *Kolumbus kam als Letzter (The Columbus-Mistake)* would be far shorter than the route through Siberia.

If we assume that there was a post–Ice Age settlement of Arctic regions a few thousand years ago, then the dry floors of the North and Baring Seas must also have been settled. The floods only came later, after the Bronze Age, as proven by megaliths on the floor of the North Sea at a depth of 50 metres. Since it is believed that the Greenland Bridge was actually frozen over, this short cut to America would have been regarded as impassable and, in the opinion of the scientists, the migration must therefore have taken place via the wide rivers of Siberia.

The post Ice Age phase with the propagated settlement of Arctic regions which were becoming free of ice is described as the Middle Stone Age, which lasted from 8000–4500 B.C. in north central Europe. If we assume that this period was a phantom time and simply rub it out, our ancestors' expansion phase would fit exactly into the post–Deluge phase I posit, around 3000 B.C. to which time, as already mentioned, the skulls of Neanderthal and Cro–Magnon Man are now also being dated. In this way, with a reduced timescale, the Neanderthal could then be the direct ancestor of the Aleut, whom he also resembles. Officially speaking, the Neanderthal maintained a nebulous presence in eastern Siberia for 100,000 years before they were suddenly transformed into modern humans 40,000 years ago, after which they strode through the supposedly frozen doors to the New World which had been in front of them for thousands and thousands of years. Let's just telescope these long periods of time and thus simplify the picture of human development and the migration of the peoples as it really happened.

Since latest analyses of Neanderthal and Cro-Magnon skulls have arrived at a maximum age of 5,000 years, their age falls directly in the Young Stone Age. For our roots cannot be seen in a single process of the Middle Stone Age (Barbujani/Bertorello, 2001, p. 23). The question of the origins of the Neanderthal however remains unanswered because we still don't know what a human adapted to an Arctic climate was doing in hot climatic regions such as Africa. Given his adaptation, he would have been out of place there and, essentially, the Neanderthal has only been proven to have been present in Europe and the (Near) East, but not in Africa. So where does the Neanderthal and, for that matter, *Homo erectus* come from?

Going by my arguments, the Ice Age is the same as the post–Deluge Snow Time, although it only started around 5,000 (or 4,500) years ago, with the winter impact after one or more celestial bodies hit the earth (cf. Darwin's Mistake, p. 202). During this post–Deluge period (= Young Stone Age), the resettlement of the almost completely depopulated European continent took place from 3000 (or possibly 2000) B.C. with people becoming sedentary and the construction of megaliths. The nomads of the tundra (Neanderthals) migrated to northern and eastern regions with the shifting of the climate zones. Did they also come from there originally and were only driven out of these areas southwards to the continental peninsula of Europe as far as the Levant because of the encroaching winter impact and Snow Time? Would this explain the sudden appearance and equally sudden disappearance of cultural layers? Are the sterile intermediate layers simply layers washed up by superfloods, which represent a short–term event and thus stand in for event–less, because non–existent phantom millennia?

Why anyway were there cultures living close to the North Pole who then migrated south? Before the dinosaur impact (K/T boundary), after shortening the Tertiary and Pleistocene around 5,500 years ago, as outlined in this book – the North Pole was free of ice. How could humans escape this inferno? "As simulations show and paleobotanists confirm, the extreme North of Europe and North America escaped the worst depredations" (*SpW*, February 2005, p. 54). Natural catastrophes would have almost completely wiped out the eco–systems in Europe and North America but the Arctic remained relatively unharmed.

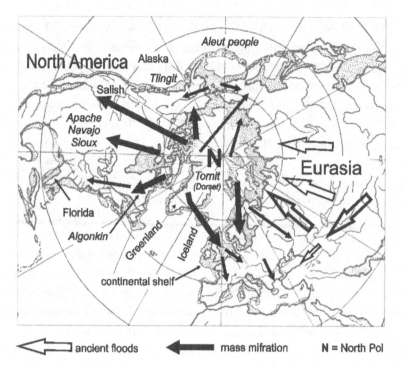

prevailed in today's Arctic regions and the water in the North Polar Sea had a temperature of 20°C. With the onset of a colder climate, a migrational movement south (North America and Europe) began. Other Hyperborean's remained and adapted to the cold. Eventually, these people too (Neanderthals and Cro-Magnon) were driven south. Once the climate improved, these tribes again moved north with their reindeer, and are now known as Inuit (Eskimos). There, they once again met up with their cousins, the Na-Dené people. The hollow arrows symbolise superfloods in Asia. This explains why the languages of Eastern and Western Siberia are related to each other but not to the language of Central Siberia.

A new study by the University of Cambridge which analysed teeth and bones comes to the conclusion that the lifestyle of the Neanderthal was similar to that of the indigenous peoples of northern Alaska. 30,000 years ago, winter temperatures are officially said to have dropped to –10°C. "Neanderthals and Aurignacian people gradually withdrew to a few fallback areas in southwest France and the Black Sea coast (*dpa*, from *New Scientist*, Ed. 2431, 24 Jan 2004, p. 10).

In any event, the "wiping out" of the Neanderthals by modern Aurignacian peoples did not take place. This is a misinterpretation based on archaeological finds which had been misunderstood. As nomads, hunters and gatherers, the Neanderthals did not become extinct, but migrated southwards only after temperatures plummeted.

The Neanderthals were forced to withdraw to the southern edge of the east European plains, when the last Ice Age began. They sought refuge on the Crimean Peninsula and the northern slopes of the Caucasus (Arsuaga, 2003, p. 323). Neanderthals have been found beyond 50° northern latitude, as illustrated by sites at Rikhta, Zhitomir and Khotylevo. "There is no doubt that they could adapt to very extreme conditions (ibid., p. 323).

Accordingly, Neanderthals and Cro-Magnon Man (Aurignacians) appeared suddenly, sometimes individually, sometimes together and sometimes – as mentioned – in seeming "reverse order". The Neanderthals live on today in their descendants in the Arctic.

With increasing temperatures after the Ice Age (= greenhouse effect after the Deluge) the Neanderthals once again moved northwards, following the reindeer north (northern Europe, Greenland, Canada) and to western Siberia. The origins of those invaders now known as Inuit (Eskimo) would otherwise be completely lost in the depths of time. At the edge of the Arctic regions, the Inuit met their American cousins, who belonged to the Na-Dené language family (e.g. Tlingit in Alaska) and the Algonquin language group (e.g. Abenaki, Micmac) in eastern Canada. The Eskimo-Aleutian and Na-Dené language families allegedly originated in the Old World. Is it possible that this is not the case and that the unacknowledged roots of all these languages lie in the North?

At the Yana River, at a latitude of 71° north, several hundred stone tips and a few spear shafts made of rhinoceros horn and mammoth tusk were discovered which resembled been assumed that the settlement of the regions above the Polar Circle took place only after the Ice Age. The newly discovered find site in eastern Siberia, said to be 31,000 years old, is twice as old as the oldest known Arctic site (*Science*, Vol. 303, p. 52).

Conventional teaching contests this view, saying that Europe's conquest took place over several phases, with four main movements, from east to west: first by *Homo erectus* around 800,000 years ago, then by the Neanderthals 120,000 years ago, followed by

early–modern Man 40,000 years ago at most and finally after the Ice Age with the development of agriculture, 10,000–8,000 years ago. The idea that early–modern Man followed the Neanderthals right across Europe using exactly the same route, in order to then completely "wipe them out" and take over their living space, is completely implausible. Population density in Europe at that time was still very low. Overpopulation was certainly not an issue. The post–Ice Age (= post–Deluge) settlement of Europe is connected with the so-called Neolithic Revolution, which was characterised by the transformation of lifestyle from nomadic to farming.

Fig. 34: Ready for Battle. *A warrior of Tlingit people, who now live on the Alaskan coast and belong to the Na-Dené - and thus are part of a global family group. From "Harper's Weekly", 1869.*

The sudden appearance of ceramics, polished stone tools, domesticated animals and cultivated plants is also mentioned in this connection, although scientists are increasingly of the view that a gradual process of change took place and the word "revolution" is being avoided more and more. Europe's settlement by Young Stone Age (Neolithic) farmers is said to have taken place from the central East, e.g. from Anatolia (Turkey) and the areas north of the Black Sea through the pre–Scythian Kurgan culture.

The oldest evidence of farming can be found in the sickle–shaped region from Palestine to the Persian Gulf which is known as the fertile half–moon. It is no co–incidence that Old Stone Age urban structures such as Jericho and Çatal Hueyuek, which were settled for long periods and rebuilt several times, are found here.

In the region of the fertile half–moon the climatic conditions were ideal as it was a winter rain region, and this is why it is believed that it was here that the evolutionary transition to a sedentary lifestyle and agriculture took place. With the onset of the younger Dryers Period (cold period) and the associated colder and dryer climate, food sources dried up. The people left places such as Jericho and other settlements. At the same time, the Ukrainian and south Russian plains were once again transformed into bare steppe landscapes (Pitman/Ryan, 1999, p. 324). The younger Dryas Period lasting 1,000 years is supposed to have ended around 11,400 years ago and, in my deluge model, should be considered the equivalent of the post–Deluge winter impact. Temperatures dropped and in southwest Asia, Europe and Africa there was barely any precipitation. As a result, the global desert belt formed from Asia across the Arab Peninsula through to Africa as connected zones (Gobi, Arab desert, Sahara).

Judging by the rock paintings, cattle were once common in the Sahara; climatic reasons made it necessary to move them south at latest three thousand years ago, confirms the geneticist, Luigi Luca Cavalli-Sforza (1999, p. 138).

In Europe's high mountains glaciers increasingly formed while from the edge of Ireland's continental shelf through to southern France and the Lower Rhine tropical temperatures prevailed, as shown by new deep drilling in Bergisch-Gladbach near Cologne in Germany. At this time the fantastic cave paintings of southern France were created, which depict naked hunters with puma, lion, antelope and other tropical animals. At the time, the North Sea and the English Channel were dry land. Great Britain and Ireland were joined to continental Europe.

The Gulf Stream was blocked by low undersea swells between England, Iceland and Greenland and diverted south along the French Coast and the Iberian Peninsula. The warm water was pressed into the Gulf of Biscay and created a tropical–warm climate here. At this time, even hippopotamus were native to central Europe whilst, at the same time, the great mountains of the Alps and Pyrenees and in Scandinavia and Greenland were freezing over. In Europe, this period represents the Old Stone Age, characterised by hunter–gatherers and nomads which did not, however, end 10,000 years ago but represents only a relatively short phase of something more than 4,000 years. In my book

The Columbus-Mistake (*Kolumbus kam als Letzter*, 2004, p. 289ff.) this briefly outlined scenario is described in detail and explained.

As there was hardly any rain in the Black Sea region either, the water supply became ever scarcer so that more water was lost through surface evaporation that was produced by precipitation and rivers. This turned the Black Sea into a non–draining lake.

But the agricultural cultures in Anatolia and in the Fertile Half-Moon were also partially or wholly abandoned by their inhabitants. Many groups retreated to areas where there was still water, in the shore areas of those rivers which had not yet dried out and the Black Sea (Pitman/Ryan, 1999, p. 325).

Some hunter–gatherer and herding groups adopted the practice of agriculture from their neighbours and became accustomed to this lifestyle. Trade flourished and in this time of need, man learned how to divert the water into the deltas, thus inventing the technique of artificial irrigation.

The Black Sea was one of the last freshwater reservoirs and it is for this reason only that people settled on its shores. The areas around the Mediterranean were plagued by extreme aridity because of the deserts forming in Africa and the Arabian Peninsula, and were therefore almost entirely depopulated. The situation afterwards improved and rainfall recommenced. Some of the inhabitants of the lake shore around the Black Sea who had been displaced by the slowing rising water level, returned to the abandoned settlements to the south (Anatolia, the Levant).

The existing remains of buildings, some of which were destroyed by earthquake, which one still finds often on the west coast of the Mediterranean, were quickly rebuilt in a haphazard way with any building material available. Sometimes neighbouring ruins were used as a quarry, which is why in ancient cities one may find that building rubble has been carelessly piled or cemented badly and sometimes back-to-front, onto well-built foundations of high quality. This destruction of settlements is usually wrongly attributed to the consequences of war.

In the meantime the sea level had risen so far that the water extended to the edge of the wall at the entrance to the Bosporus Valley and then poured into the Black Sea which was 150 metres lower. Around 50 billion cubic metres of saltwater poured every day into the former freshwater lake and destroyed the entire fish population, until finally new saltwater fish from the Mediterranean became native there.

The Young Stone Age agricultural settlements of which we are aware were exclusively in water–retaining marshy ground, as the Russians have discovered during their sediment drillings everywhere on the broad continental shelf of Bulgaria, Romania, the Ukraine and Russia, which is now above water. The farmers, whose settlements can now be found deep below the water level of the Black Sea, had to flee from the encroaching waters. The peoples living around the Black Sea were scattered in all directions to the west (central Europe), the north (eastern Europe), the east (Asia, as far as Japan) and the south (the Levant, Mesopotamia and Egypt) (cf. Pitman/Ryan, 1999, p. 325ff.).

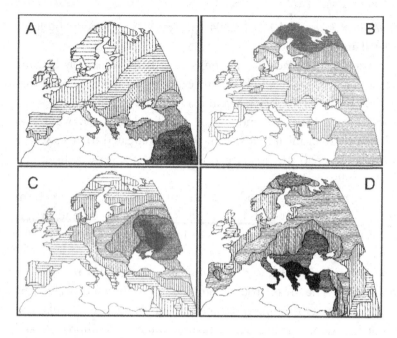

Fig. 35: Gene. According to Cavalli-Sforza (et al. 1994), the first four main components of Europe's genetic map result in landscapes.

Map A documents the westward expansion of the Neolithic tillers of land, which led to sparse settlement.

Map B shows a genetic adaptation to the cold from north to south "probably as a result of a single big migration"– Neanderthals and Aurignaciens. The Megalithic Period with Cro-Magnon man, including the Basques, whose Vasconian original language is evident from Central Europe (Bavaria) to the Black Sea, also belongs in this phase.

Map C shows westward (Europe) and southern (India) military expansion in line with the climate improvement, which started from the Kurgan region (including the Scythians) on the Black Sea. Map D shows the Greek expansion in Europe. Core areas are dark. From: Luca and Fracesco Cavalli-Sforza, 1994.

In Europe the areas left vacant by the retreating hunting and gathering peoples were occupied by the military invaders of the Kurgan people (Scythians) moving westwards from the Black Sea (cf. Renfrew, 2004, p. 44). They brought their longhouse architecture – which was later adopted by the Celts, Germans and Vikings – their artisan skills (ceramics) and their farming methods to their new (but simultaneously old) home. A second expansion started from Greece in the 2nd millennium B.C. (cf. Cavalli-Sforza, 1996, p. 248). These expansions (Maps C and D of Fig. 35) are discussed in depth in my book *The Columbus-Mistake*.

The Scythians and their forerunners, the Kurgan culture, lived north of the Black Sea up to the Caspian Sea. The horse, which the Kurgan and Scythians used for riding, was domesticated here. These equestrian peoples, who are often regarded as Mongolians, were not of inner–Asian origin but Indo–Europeans, who expanded westwards to Greece, Europe, Great Britain and Scandinavia (Renfrew, 2004, p. 44). These peoples also brought their language to Europe out of which Gothic, Celtic and "Teutsch" emerged, as Johann August Egenolff described in his book *Historie der Teutschen* Sprache

(History of the German Language) which appeared in 1735. The megalith-building peoples "were therefore Indo-Europeans, out of whose language today's Celtic languages developed. As such, Europe's entire prehistory appears as a serious of constant migrations … on a common early Indo–European basis, with a few non-Indo-European remnants. The impetus of this development was not a series of migrations but the complex interdependencies in Europe with its largely agriculturally-driven economy and Indo–European language" (ibid, p. 47).

Changes in Habitat

This short sketch of history after the Deluge (or alleged Ice Age) clearly illustrates that the regions of Europe were inhabited by only a few hunters and gatherers and were settled only relatively late in the Young Stone Age. This doesn't mean that there were no cultures before this in Europe and the North. Officially no thought is given to the fact that today's Arctic regions were settled by sedentary peoples during the "Ice Age.

In ancient times, these settlers were a mythical and unreachable people far to the North at the edge of the world, called Hyperborean's. Later, this name came to mean all people in the extreme North, as every effort to find them failed because they had fled across land and sea in the direction of the Mediterranean and the Black Sea to escape huge floods and drops in temperature. What remained was an Arctic which was starting to freeze (Canada, Greenland, Spitsbergen) and an almost completely depopulated Europe, with a radically different shape to that with which we are familiar today.

Several dozen once large lakes no longer exist today, such as the Upper Dnieper Lake, the Upper Volga Lake, Lake Tunguska, Lake Pur or Lake Mansi (Grosswald, 1980, Fig. 7). The map of the earth drawn by the Greek cartographer, Eratosthenes in the 3rd Century B.C. shows a connection between the Caspian Sea and the Arctic Ocean. Today, they are separated by 2,200 km of land. Did Eratosthenes just dream it up? But the geographer, Pomponius Mela also published a map, showing that the Caspian Sea in the Land of Scythians was connected by a wide river to the Arctic Ocean (there known as the Scythian Ocean to the north (Fig. 36, p. 159). These maps which already existed in ancient times may go back to the pre–Greek culture of the Carthaginians (Hapgood, 1966, p. 113).

The presence of the Caspian seal lion (*Phoca caspica*) which resembles the Baikal seal, in the northern part of the Caspian Sea supports this connection to the Arctic Ocean. These animals' access to the Arctic Ocean was cut off by the falling level of the Caspian. Geological facts also seem to support the connection. A large flatland region, the Caspian Depression, with an area of 200,000 km² and a slight inclination in the direction of the Barents Sea (North Polar Sea) indicates the erstwhile connection. The Caspian Sea fills the deepest point of a depression on the earth's surface, which extends up to 28 metres

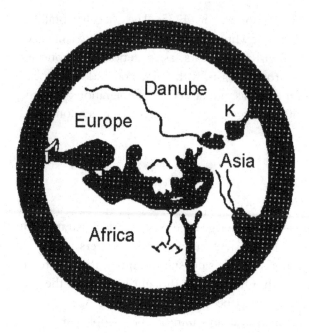

Fig. 36: Flat Earth. The antique world view at the time of Herodotus (490–430 B.C.) shows a flat earth surrounded by water, connected to the Caspian Sea (C). To the north, the earth ended with the tribes of the Kimmerians, a people related to the Scythians. These people, already mentioned by Homer, lived at the entrance to Hades. Their lands were in perpetual darkness (Kimmerian darkness) and Helios's light did not shine here. The legendary Hyperborians - the "people beyond the North" lived somewhere to the north or northwest.

beneath present sea level. This Caspian Depression formed a large lake with a much higher water level after the Deluge and was connected with the Barents Sea and thus the North Polar Sea as well as the Atlantic, the Bering Sea and the Pacific.

According to a study by P. A. Carlin (et al., 2002, pp. 5–6 and 17–35), several lakes in Siberia which no longer exist were filled by megafloods from the mountains of Asia at the end of the "Ice Age" up to 13,000 years ago (= 4,500 years ago after the Snow Time). The water finally flowed into the Caspian Depression and drained from there into the Black Sea. This study (Carlin et al., 2002) however assumed that the draining of the Caspian Sea into the North Polar Sea was blocked by the alleged glaciers of the "Ice Age". If we however take my Snow Time model rather than the Ice Age model, there was indeed heavy snowfall and ice formation, but there were no kilometre-high glaciers in the flatlands but only thick ice sheets on the high mountains (cf. *Mistake Earth Science*, p. 184ff. and 248.). Taking the Snow Time Model, the Caspian Sea could have drained northwards into the North Polar Sea (Photo 65) and the latest study of superfloods in Asia complements and even underpins the depictions in the old maps.

The precondition for this scenario is superfloods, discussed in my books and strictly rejected by conventional teaching up to now. But this viewpoint, which I have been discussing since 1998, has now been explicitly confirmed by the journal *Science* (29 Mar 2002, Vol. 295, pp. 2379–2380): the great basins with "Ice Age" lakes in Siberia (incl. the Caspian and Aral Seas) are the witnesses of these floods, which even flooded broad highlands which were in the way, hundreds of kilometres away. They caused grooves to be cut into the mountains which can be clearly seen in satellite images of central Asia. The enormous Takla Makan Desert (Tarim Basin) lies in western China. Here, there was

Debris flow

Gravel

Fig. 37: Superfloods. *In only a few hours, a superflood bursting from the Altai Mountains in the direction of the Caspian See left huge gravel deposits in the little Yalomon Valley (Eastern Kazakhstan).*

once a great lake, comparable to the Caspian Sea, along whose northern shore the Silk Road, an old trading route, ran.

As already mentioned, a new study based on skull analysis suggests that the Neanderthals migrated through Siberia to North America. Was the Neanderthals' or pre-Eskimos' route via the waterway shown on the old maps between the Caspian and the Barents Sea (North Polar Sea) and through the many wide rivers and large lakes of Siberia not blocked as a result of superfloods from Asia? Yes! For the Neanderthals who were adapted to the Arctic cold came originally from the north, but not from the south or the east (Siberia). They (the Hyperborean's?) were displaced southwards to the continental peninsula of Europe by the continually worsening climate in the north, brought about by the post-Deluge Snow Time. Researchers working with Tjeerd van Andel at the *British University of Cambridge* (*New Scientist*, Ed. 2431, 24 Jan 2004, p. 10) emphasise that the migrationary movement of the Neanderthals and the Aurignacian peoples southwards coincides exactly with the advance of the ice sheet in northern Europe.

These "Old Stone Age peoples" pushed forward to a few areas of refuge in southwest France and on the coast of the Black Sea. This mention of the flight of modern humans from central and northern Europe to areas around the Black Sea, before the subsequent re–settlement of Europe during the period known as the Neolithic Revolution, corresponds with the displacement I described in my book *The Columbus-Mistake* of central and northern European peoples from areas which, like the erstwhile steppe in the area of today's North Sea, were flooded. At this time, England and Ireland were still connected to the European mainland. This flooding and separation of Great Britain and Ireland from continental Europe went hand in hand with a drastic change of climate, accompanied by intense tectonic activity (cf. Hsü, 2000, p. 174).

In the Baltic too, there was also a Deluge which by official dating took place only just over 6,000 years ago, leading to a rising of the sea level and a dropping of the earth's crust. Under the aegis of the DFG (German Research Foundation), archaeologists on the "Sincos" (sinking coasts) project have discovered several Stone Age settlements and the remains of sunken forests, in depths of up to 7 metres on the floor of the Baltic Sea in the Bay of Wismar. Further find sites, for example, before the Baltic island of Rügen, are being examined. During the period 5400–4100 B.C. Stone Age hunter-gatherers, the Eteboelle people, eponymously named after the Danish find site at Limfjord, lived in the Baltic Sea area. They dwelt in small villages, undertook maritime voyages and lived from catching fish and hunting. They were present from the mouth of the Elbe through to Denmark, southern Sweden and Poland. From an anatomical point of view they appear archaic. In the 1930s the skull of one of these hunter–gatherers was discovered during construction work on the Rügen Dam. The skullcap was a good centimetre thick and displayed a prominent browline – an indication of robust characteristics of the Neanderthal type? Subsequent examination found that, in addition the man had been scalped Indian–style …

Is it because of this flood catastrophe that the Danes are not able to definitively relate their cultural history? They are missing an important element between the farming culture in the flat mountain regions and the Danish hunter-gatherer culture. Was northern Europe's cultural past buried under the floods of the formerly settled Baltic and North Seas? "Mecklenburg has only been on the Baltic for 7,000 years" and "the level of the southern Baltic rose, just like the water in a bathtub which is lifted on one side … Before this process was evident in northern Germany, the Baltic Sea extended roughly to a line which is now at least 25 metres beneath sea level", said Prof. Dr. Kurd von Bülow (1952, p. 49). In Fig. 38 one may see that because of the Darsser Schwelle (ridge) the actual Baltic Sea was separated from the Belt Sea which was largely dry, apart from the original river valley. The North and Baltic Seas were *not connected* to each other at that time. The Baltic Sea was formerly an inland lake and today it is still the biggest brackish water lake in the world.

An essential part of our history now lies buried beneath the waves of the Baltic and North Sea. Should the Baltic Sea by the melting, supposedly several kilometers high glaciers of the "Ice Age" in Scandinavia and northern Germany have not been filled to the brim with melt water? The "Ice Age" theory is incompatible with the "post-Ice Age" settlement of the dry floor of the Baltic Sea, which was filling up very quickly with water from superfloods coming from Asia, which then drained away again via the Caspian into not only the North Polar Sea but also into the eastern Baltic (Finnish Lakeland) and across the ancient river valleys into the western Baltic (Belt Sea) (see Photo 65). The official view (Wahnschaffe, 1921) of the ancient river valleys of northern Germany is that they were formed "immediately after the Ice Age, with a greater volume of water" (Dacqué, 1930, p. 62).

There was a cultural connection with the regions far to the east: "In terms of craftsmanship Siberia is without doubt an extension of sub-Arctic Europe" (Müller-

Fig. 38: Flooding. The Darsser Schwelle (ridge) separates the Belt Sea to the west (left) from the actual Baltic. The Belt Sea is no deeper than around 25 metres while the swell is 18 metres; the Baltic falls eastwards to greater depths. The floor of the Baltic (white areas and more) was once inhabited. From Bülow, 1952.

Beck, 1967, p. 391). It is as good as certain that the first stage of the Siberian Old Stone Age is nothing more than an extension of the outgoing Young Stone Age in eastern Europe, despite the distance and the absence of intermediate stages (Chard, 1958).

Today, a considerable part of our prehistory lies buried beneath the waves of the Baltic and North Seas. Shouldn't the Baltic have been practically overflowing with meltwater from the melting glaciers of the "Ice Age" in Scandinavia and northern Germany, said to be several kilometres high? The Ice Age theory is incompatible with the "post-Ice Age" settlement of the dry floor of the Baltic. In fact, the Baltic filled very quickly with water from superfloods coming from Asia, which then drained away again via the Caspian into not only the North Polar Sea but also into the eastern Baltic (Finnish Lakeland) and across the ancient river valleys into the western Baltic (Belt Sea) (see Photo 65). The official view (Wahnschaffe, 1921) of the ancient river valleys of northern Germany is that they were formed "immediately after the Ice Age, with a greater volume of water" (Dacqué, 1930, p. 62).

There was a cultural connection with the regions far to the east: "In terms of craftsmanship Siberia is without doubt an extension of sub–Arctic Europe" (Müller–Beck, 1967, p. 391). It is as good as certain that the first stage of the Siberian Old Stone Age is nothing more than an extension of the outgoing Young Stone Age in eastern Europe, despite the distance and the absence of intermediate stages (Chard, 1958).

Nevertheless the missing intermediate stages in Siberia should point us to a different interpretation. On the one hand, the Old Stone Age evidence in western Siberia was imported from eastern Europe and/or via Scandinavia but on the other hand these finds are absent in central Siberia. Correspondingly distant finds in eastern Siberia therefore indicate that the settlement direction is reversed, from east to west, i.e. from America across to the dry Bering Strait to eastern Siberia, but not stretching as far as central Siberia. If the people of the Old Stone Age originally came from northern regions, the routes from there to Europe on the one hand, via the Greenland Bridge and, on the other to eastern Siberia via Canada are equally far. That would mean an Old Stone Age in the Arctic! There are indeed unmistakable Stone Age artefacts and correspondingly ancient rock drawings in hunter and fisherman style in settlements north of the 70° latitude on the coastal strips on both sides of the North Cape in Scandinavia. The sets of implements display forms from all three stages of the younger Old Stone Age, whereby the typical Aurignacian is predominant (Nummedal, 1929, pp. 92, 95, 97ff.). Old Stone Age implements of modern humans from the Aurignacian should really be at least 30,000 years old. Conventional teaching says that at this time, the landscape north of the 70° latitude was covered with glaciers several kilometres high and would therefore have been completely inaccessible.

But it is not only in northern Norway that one finds rock drawings, but also as far as 700 km further north on Spitsbergen close to the 80° latitude (cf. Photo 69)! This is naturalistic and indisputably Old Stone Age rock art, although there are only Russian reports about them which provide little information (Simonsen, 1974, p. 132f.).

This scenario does not fit into the prevailing assumption that a settlement of Scandinavia took place only 10,000 years after the end of the "Ice Age" and the retreat of the giant glaciers. The Norwegian newspaper Aftenposten reported in 1997, that Scandinavia's settlement began far earlier than had previously been thought (*BdW*, 22 Oct 1997). Paleoanthropologists have always found it difficult to interpret the Old Stone Age finds in northern Norway mentioned above as these cannot be reconciled with kilometre-high Ice Age glaciers in Scandinavia, particularly since old settlements in the area of the "Ice Age" glaciers have been found – a riddle for researchers. Reducing the length of human history and shifting the younger Old Stone Age into an ice-free post-Deluge phase solves this apparent riddle. Only during and afterwards did the allegedly very long-lasting Ice Age take place in the form of a relatively short-term Snow Time: while Old Stone Age people in central Europe were still hunting tropical animals naked, as seen in the cave paintings in southern France, the high mountains started to freeze over, in Scandinavia as well, followed by north and central Europe, which were covered with a (thin) shroud. The tropical animals, including hippopotamus, water buffalo, elephant, tiger, apes (*Macaca sylvana suevica* in Heppenloch, Germany), rhinoceros and lion – as well as humans (Neanderthal and early-modern Man) died in great numbers here in Europe. They were also destroyed, crushed and washed together into gigantic deposits by raging storms and floods (cf. *Mistake Earth Science*, p. 179ff.). This is why the remains of modern humans are found in apparently anomalously old layers.

Following these catastrophic events, Europe was laid waste and depopulated. The old cultures were destroyed, many peoples fled south in the face of the encroaching cold while others – both humans and animals – crawled into caves where they froze to death. Because of where they were found, many animal skeletons discovered in caves were interpreted as belonging to cave-dwelling animals – e.g. lions as cave lions – which was a mistake. There were even supposed to have been cave hyenas in Germany. There were, without doubt, hyenas in central Europe, (as there were in Africa) but these animals lived not in caves but as they do today in Africa: moving freely across the sweeping savannas of Eurasia, like the Saiga Antelope and the mammoth.

With the shifting of the climate zones southwards the squat nomads adapted to the Arctic (Neanderthal, Eskimo) were displaced from northerly areas to America and the European peninsula. Here, the once animal-filled steppe had been transformed into a bare tundra landscape. The nomads followed the reindeer, who also lived in the Rhine Valley. Did the origin of the cultures lie in the north rather than the south? At this time there was much more land in the north than today. The Barents Sea north of Scandinavia and the North Sea are relatively flat and when the water level was low, there were steppe landscapes here (cf. Photo 69). Before the "Ice Age", the sea level is said to have been considerably lower: at least 130 metres (*Science*, 1979, Vol. 204, pp. 618–620).

It remains highly improbable that groups of people advanced voluntarily from the south into the inhospitable latitudes at the edges of the glaciers. It is more likely that people adapted to the cold evaded the cooling climate in the north and followed the animals southwards, to then return north when the climate became warmer. Reindeer skeletons have been found in Hamburg's Tunnel Valley – in the triangle between Hamburg-Meiendorf, Ahrensburg and Stapelfeld. This valley bears witness to great masses of water, since at its narrowest point between the mountains *Harburger Bergen* and the hills *Blankeneser Höhen* it is still 8 kilometres wide! At the time, the water flowed on into the Elbe river valley, from where huge quantities of meltwater from the allegedly thawing northern inland ice flowed into the North Sea. These were in fact waters both from floods, which took place during the Alluvium (Holocene) also leaving many erratic blocks behind and from superfloods originating in Asia.

Traces of Stone Age Man have been found in this area close to Meiendorf. They were deemed to be reindeer hunters of the so-called Hamburg Culture. The official view is that post-Ice Age, 12,500 years ago, this was a treeless tundra, which only became wooded again 2,000 years later. For this reason, the Old Stone Age reindeer hunters moved away again with the change in climate northwards and eastwards to colder tundra regions. Around 10,200 years ago, after a new cold period (younger Dryas Period or Dryas III), the reindeer hunters supposedly returned with improved hunting techniques (bow and arrow). This was the so-called Ahrensburg Culture, which has also been found on the southern edge of Cologne Bay (Floss, 1989) in the central German mountains (Baales, 1996) and in the Belgian Ardennes (Baales, 1999).

Numerous crushed bone fragments show that the reindeer hunters even extracted the fat in the bones by boiling these shards. The bone grease obtained in this way was processed with berries and dried meat into a long–lasting food reserve, similar to the pemmican produced by the North American Indians.

A reindeer breeder in northern latitudes requires large areas for grazing, in order to provide enough food for his herd of up to 5,000 animals. Are there major differences between the reindeer breeders (possibly reindeer hunters originally?) still living there today and the Ahrensburg Culture? Did they already have herds and did the Lapps live just like their ancestors until recently? Although genetically European, the Lapps have genes which differ most from those of other Europeans (CavalliSforza, 1999, p. 132).

The tundra landscapes in Europe extended from the north German flatlands and the adjacent lowlands of northern Belgium and the Netherlands as far as southern France to the west and the Baltic States in the east. Reindeer have been found in all of these regions, including in the northern Central German mountains. The origin of reindeer (caribou) is unknown. "When they appeared in Europe they were already fully adapted to their Arctic habitat" (Paturi, 1996, p. 429), so they appeared out of the blue.

In 1932, Alfred Rust undertook excavations (Rust, 1937) in Meiendorf in the Hamburg Tunnel Valley previously mentioned, and found in the so-called Hamburg Culture not the reindeer known from Siberia (*Rangifer tarandus*) but, to the surprise of the excavators, a different type of reindeer, *Rangifer arcticus* (Gripp, 1937, p. 72). The surprise was in the incidence of *Rangifer arcticus* (today also known as *Rangifer tarandus arcticus*). In Germany *Rangifer arcticus* was found above the main layer of the Mosbach Sands in Wiesbaden (Probst, 1999, p. 303). The inhabitants of various habitats such as elephant, bison, giant deer, hyena, beaver, bear, rhinoceros, reindeer and even hippopotamus have been found in the widespread Mosbach layers which are over 900,000 years old (Brüning, 1980).

However, the *Rangifer arcticus* found near Hamburg came from the younger Old Stone Age and is 10,000 years old at most. What is surprising is today's incidence of this caribou species, because *Rangifer arcticus* lives in polar North America from the Hudson Bay to Alaska and can also take in a piece of (East) Siberia. Should one see this as a clue to Old Stone Age finds in East Siberia? Did the reindeer hunters follow the caribou and cross the then dry Bering Strait in a westward rather than an eastward direction? Did the people follow the caribou with the shifting of the climate zones from Arctic regions southwards up to the European continental peninsula? After all, the finds of *Rangifer arcticus* in northwest Eurasia and East Siberia are separated by 80° of longitude. Would this also explain why there is a language vacuum between the Eskimo-Aleutian languages in East Siberia and the related *Chukchik-Kamchadalic* languages in West Siberia, which is now filled by a Turkic language (Altaic) which originated in central Asia (Photo 63)?

In his zoological monograph, Arnold Jacobi defends his theory that *Rangifer arcticus* came directly to Northwest Eurasia from North America and he cites Alfred Wegener's theory of continental shifting. Wegener himself believed in a land bridge between the

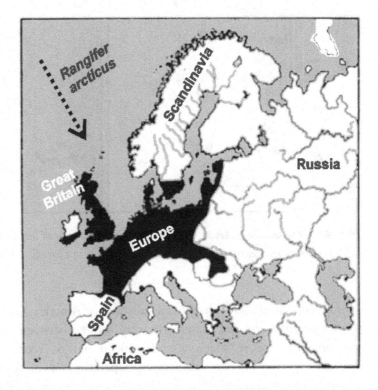

Fig. 39: Distribution. Rangifer arcticus in north-western Eurasia (after Jacobi, 1931). These reindeer, whose origins are officially unknown, came across the Greenland Bridge.

the Atlantic western part (North America) and the eastern part (Europe) nor in the older Quaternary, i.e. at the start of the "Great Ice Age", at least at the most northerly point of the ocean. Just before the rift period are the two continental shelfs supposed to have separated (Jacobi, 1931, p. 40ff.). This is why small creatures, such as two species of the beetle *Bembidion grapei Gyll*, who only survived close to hot springs, have been found embedded in advancing ice in south Greenland and Iceland (Lindroth, 1957, p. 277f.). Carl H. Lindroth collated further examples of (amphiatlantic) fauna as proof of a land bridge between north America and Europe: two species of bird as well as species of butterfly, moth, spider, snail and beetle. The garden snail (Cepaea hortensis) too was discovered in a prehistoric heap of seashells in America (Lindroth, 1957, p. 234).

This confirms that there was a land bridge between Europe and North America, which I discussed in detail as the Greenland Bridge in my book *The Columbus-Mistake*. According to Wegener, at least part of the continental shift took place at a time when *Homo erectus* and the Neanderthal populated Europe. I share this opinion and would add that Fridtjof Nansen's 1893–1896 polar expedition with his ship Fram proved that the greater part of today's deep-sea floor in the north Polar region, which now lies at a depth of 1,000–2,500 metres between the volcanic Jan Mayen Island and Iceland, only "dropped in very recent times by 2,000 metres" (Walther, 1908, p. 516).

Iceland is made up of the mountains of a sunken island, whose once dry valleys today are now deep fjords like those in Norway. When this landmass sank, far-reaching changes took place in the lithosphere. "This strongly suggests that hand in hand with this, a significantly different distribution of the masses must have taken place, which would not have been without influence on the earth axis" (Walther, 1908, p. 516).

163

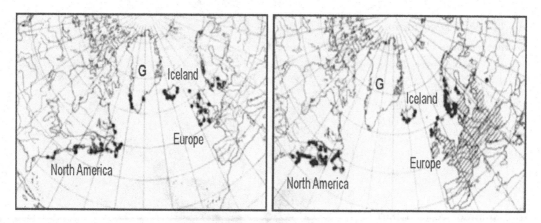

Fig. 40. Transatlantic Areas of Distribution. *Examples of amphiatlantic fauna. Left picture: garden snail. Right picture: extinct great auk (Penguinis impennis) G = Greenland. From Lindroth, 1957.*

As a result, the adjacent continental substrata of America, Greenland and Europe moved somewhat apart and new waterways were formed between them. According to modern tectonic plate theory which was developed out of Wegener's theory of continental drifting, these continents have been separated for at least 65,000,000 years – an unsurmountable obstacle for caribou. I have, however, already mentioned that the Tertiary is a phantom period: with that in mind, the Greenland Bridge slides into the late Quaternary (Diluvium).

Thus the way was for free for *Rangifer arcticus* and the hunters with him (Eskimo, Neanderthal) to move in a southerly direction towards Europe. The eastern route to Siberia was, after all, also blocked by superfloods and the post-Deluge (post-Ice Age) waterway between the Caspian and the North Polar Sea and by the wide, north-flowing torrential rivers of Siberia. In addition, the route across the Greenland Bridge was not only available, but was also free of ice and far shorter.

If the ceramics from Northwest Russia are more closely related to those of North America than with those of East Siberia or with Baikal Ware (Ridley in Pennsylvania Archaeologist, 1960, p. 46ff.), then it is not surprising that blonde, fair-skinned Eskimo types with Nordic appearance and beards have been recorded on the northern coasts of East Canada, east and north of Hudson Bay and on Greenland, who differ considerably in appearance from their mongoloid countrymen (Greely, National Geographic Magazine, Vol. XXII, No. 12, December 1912; cf. Stefansson, 1913).

The Dutch scholar, C.C. Ulhenbeck had already tried to establish the connection between Greenlandic belonging to the eastern group of Eskimo languages (and the closely related idiom spoken on the Labrador Peninsula) and Indo-Germanic (Jensen, 1936, p. 151). Today, the Eskimo-Aleutian languages are classified as Euro-Asiatic, as is Indo-Germanic (Greenberg/Ruhlen, 2004, p. 63).

Contrary to plate tectonic hypothesis, there was a transatlantic land bridge in the late Quaternary. This fact led the discoverer of the reindeer skeleton at Meiendorf, Karl Gripp to the conclusion, disguised as a question, that Old Stone Age Hamburg hunters of the Magdalenian culture might be related to the Eskimos, given that both had a highly developed sense of art (e.g. throwing sticks) (Gripp, 1937, p. 72)? Behind this question was the assumption that the diluvian reindeer hunters went, like the reindeer themselves, southeast from the Canadian Polar Regions. These facts fit best with the assumption that, from time to time, depending on climatic conditions at some point outside Europe, roaming outsiders and their animals came to Europe. This would also explain the appearance in Europe, out of the blue, of "hunting groups from the younger Old Stone Age with Eskimo lifestyle" (Rust, 1962, p. 63).

If one bears the equipment of this people in mind, the Eskimo lifestyle has, according to latest research, lasted from at least the beginning of the Young Stone Age up to today. These hunters have been children of the Arctic since the Aurignacian to Magdalenian cultural phases. Their stature, their fur apparel with boots, their house-shaped leather tents with heatable rooms and their special relationship with fire indicate that they grew up in wintry regions.

For example, it has been shown by examples in Ostrau-Peterhofen at the Moravian Gate, on Spitsbergen and in west Greenland, that the people of the younger Old Stone Age even used coal for their fires if a seam was accessible.

The scholars of the 19th Century, not yet intoxicated and befuddled by the drug of evolution theory, saw the transatlantic connections objectively. On the basis of his anthropological studies in Brazil, Paul Ehrenreich summarised:

"We know that in more recent geological periods, Asia and Europe were connected with North America. A circumpolar landmass existed at a time when, if not Man himself, then his last ancestor, inhabited the northern hemisphere. There is … therefore not the least reason to assume that America was uninhabited at this time, since Asia or Europe already had a population" (Ehrenreich, 1897, p. 42). The landmasses encircling the North Polar Sea are, after all, shown on old maps.

Since the postulated chain of human evolution from Australopithecus via Neanderthal to modern Man could not be proven in America, the first Americans must have come from

Fig. 41: Eskimos from Labrador. Nordic-European appearance of the Eskimos. From Rakel, 1894, p.725.

Fig. 42: Cold Protection. The fully intact skeleton of a mammoth hunter, found in Vladimir (Russia). The man lived around 35,000 years ago, was tall and wore trousers and shoes made of fur. Thus we have a different picture of primitive humans to that which is presented in many schoolbooks.

somewhere else. As the transatlantic route was apparently blocked by wide waterways, only the path across the Bering Strait was thinkable. It was said that the entire American continent down to the tip of South America was settled from here. The theory therefore originates in evolutionist thinking and was adopted without much ado by archaeology.

Nonetheless, archaeological finds in east Siberia make an export from America just as likely. Doesn't America give as well as take? This is what Alan Lyle Bryan was trying to prove in his 1978-study. The geographical extension of bifaces – two-sided projectile points (*commonly called*: arrowheads) – on Asia's north Pacific coast makes it highly probable that this was a diffusion from America, as Tolstoy (1958) suggested with regard to the late Neolithic and Bronze Age projectiles of the Baikal region. According to Bryan (1978) the artistic shapes of the wonderfully worked bifaces of the American Folsom, Clovis, Sandia and Yuma types cannot be found anywhere during this period.

Let us however take a look at the technically advanced stone tips of the North American Clovis culture as this special type is also known in Europe. There, they are assigned to the Solutrean culture which is said to have existed in the period from 22000–18000 B.C. If only these arrowheads were found in Europe, Africa or Asia, the find site would be marked on every map as a Solutrean site. But since these find sites are in America and, in the meantime, several such finds have been made, they were unanimously declared to be fakes – what can't be, shan't be!

As the bifaces of the Clovis culture are comparable to those of the Stone Age Solutrean culture in Europe, the export of technology could be documented (*Science*, Vol. 286, 19 Nov 1999, pp 1467–1468).

The archaeologist, Reid Ferring of the *University of North Texas* in Denton continues: If we assume that there was no ocean (Atlantic), we would immediately feel ourselves

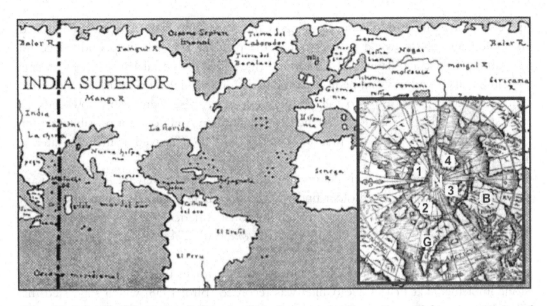

Fig. 43: Nordic Homeland. Into the 16th Century, America was regarded as an extension of Asia (Greater India), following the concept of the ancient cartographer, Claudius Ptolemy (allegedly 2nd Century). The dotted line shows Ptolemy's border between the Old and the New Worlds. In 1414, Albertin DeVirga produced the first representation of a connected landmass in the Arctic North. Until far into the 16th Century, maps showed a connected landmass in the Atlantic Region, such as in J. Gastaldi's 1548 map, shown here: the land of the Hyperborean's. Insert: other maps, such as that by Oronteus Finaeus (1532) show large islands (1 to 4) around the North Pole. The Barents Sea forms a landmass with Scandinavia (B). G = Greenland – has mountains and no ice.

drawn to Western Europe as the area from which the Clovis people originated.

The demand for a non-existent ocean or a land connection between Europe and America, but just met the Greenland Bridge. Perhaps it is the other way around and America is the home of European Solutrean culture? A seemingly heretical question, but the other, in principle, possible alternative shows.

Scholarly Indians such as Prof. Vine Deloria (1995) believe that this is the case and have documented their opinions. But perhaps both approaches are wrong and the European Solutrean as well as the American Folsom culture came originally from the north.

If on comparison of two-sided stone tools from the American Clovis with the European Solutrean culture it is felt necessary to propagate an obviously transatlantic cultural exchange, then set us the time ladder of human evolution in Europe, a trap. The Clovis culture is supposed to be only 10,500 years old which, when measured alongside old European cultures, is at least 7,500 years too late. Therefore, any suggestion of a cultural connection must be ruled out on official archaeological, paleo-anthropological and evolutionary grounds. However, going by the rejuvenation theory already mentioned, the Solutrean culture would fit better into the post-Ice Age Clovis period. Since the revelation of false age datings of Old Stone Age skulls in central Europe, new datings should be taken seriously and be used as the basis for a rethink. If we push the Solutrean culture in Europe forward into the Young Stone Age, these European arrowheads are now suddenly younger than those of the Clovis culture in America. Is it possible that this particular technique appeared first in America and then arrived somewhat later in Europe?

After great debate, the Clovis culture in America is now relatively well-established and the debate is now about even older finds. Given that there have been many comparable finds, the spread of Clovis culture throughout north America and into central America within a few centuries is now taken as fact (*Science*, Vol. 274, 13 Dec 1996, pp. 1820–1825).

It seems this culture even reached south America, for in the 1930s a chronology of stone implements at Fell's Cave (Chile) was recorded (Bird, 1938) which suggested that Chile had been settled early on (*The Geographical Review*, Vol. 1, 6/1938, pp. 250–275).

In the 1960s a fluted stone implement was discovered there (Dillehay, 2000, p. 98). The anthropologist, Thomas D. Dillehay of the *University of Kentucky* in Lexington notes that it is the fluting which worries most archaeologists because this feature is also found in the Clovis stone arrowheads in North America and one can therefore assume there was a connection between the two continents (Dillehay, 2000, p. 98).

Finds of further stone arrowheads with fluting have also been reported in Colombia, Venezuela, Ecuador, Argentina, Chile, Uruguay and southern Brazil. Thus contrary to the conventional teaching, comparable projectiles have been found not only in north America but also in south America and in Europe. Early-modern Man (Cro-Magnon) is therefore widely represented: in Africa, Europe, Asia and America as this seeming explosion of finds demonstrate. In the meantime, however, evidence of the even older presence of humans in America is being recorded. Up to now, at least the pre-Clovis sites of Meadowcroft and Cactus Hill in north America and Monte Verde in southern Chile, said to be 30,000 years old (Dillehay, 2000) are acknowledged. Other find sites are being excavated and are waiting to be officially recognised.

Once these excavation sites of American finds from the Old Stone Age have been recognised a problem will arise, because the Old Stone Age skulls in Europe will be seen to be drastically younger, whilst America's prehistory can continue to extend into the (nebulous) past. Or are the age datings of the American finds too old as well?

After all, artefacts aged up to 33,000 years old have been found in El Cedra, in the Mexican state of Sinaloa, together with the tarsal bones of elephants "in undisturbed stratified deposits". It is now known that the elephant became extinct throughout the entire American continent a long time ago (Lorenzo/Mirambell, 19986, p. 107; cf. Cremo/Thompson, 1997, p. 192).

Neanderthals in the New World

Until modern times, the custom of reshaping skulls was still common on Crete and in Lapland. Nine human skeletons from the 5th Century were found in a plundered burial field close to the Lower Austrian villages of Kronberg and Kollnbrunn. In five of the nine graves, the skulls displayed signs of man–made deformation caused by binding shortly after birth. There are similar examples of deformed long skulls in the Orient, Egypt, Nubia and, above all, in China.

In Peru's state museums in Ica and Lima, I was able to appraise several long skulls, also known as "tower skulls". Many of these skulls date back to the Paracas Culture which emerged around 800 B.C. and the subsequent Nazca Culture (around 200 B.C.). Purely coincidental parallels between South America, the Old World and China?

Broad skulls have however also been found in south America. The Neanderthal had a bigger skull than modern Man. If such a skull is deformed, it is without doubt possible to archive a long or broad shape. However, the skull with two swellings, which I was able to photograph at the Lima Museum (Peru) and a melon–shaped skull are also curious. The first shape is surely not simply achieved by binding and the other is distinct in being considerably more voluminous than a normal skull. In other words, these skulls must have belonged to bigger people because otherwise the head would not have fitted through a normal birth canal.

But this brings us to a problem which is rarely discussed, because the anthropologists seem to have no interest at all in these skulls. Because they present several unsolved problems? Certainly, some long skulls may be seen as the result of man–made deformation. Both the original and the deformed skulls have the same volume, are therefore the same size in terms of content. The biggest skull of a modern human ever documented in medical literature had a volume of 1980 cubic centimetre, but the shape of the skull was perfectly normal. However, the deformed skulls in south America include some with a volume of 2200–2500 cubic centimetre, significantly larger than the skulls of Neanderthal and Cro-Magnon Man. The unusually shaped skulls are even bigger; with a volume of around 2600–3200 cubic centimetre. This has therefore given rise to speculation that these might be the skulls of extra–terrestrials. In view of the eye cavities, which are 15–20% bigger than usual, one must assume that these are giant skulls belonging to very big people: male Neanderthals would then have to be 2.3–2.6 metres high rather than their average 1.65 metres. Are the myths and traditions which speak of giants true?

The giant skulls recorded in south America therefore represent a problem since very big people (giants) would have had difficulties in surviving under present conditions. Given today's gravity it would probably not be possible for humans to attain heights of 3 metres and more.

There is also another problem, because according to anthropological criteria, the

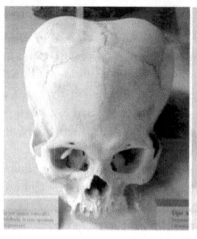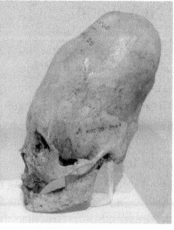

Fig. 44: Long Skull. In South America there are strange skull shapes with a volume almost double that of the biggest skulls of modern Man. The author photographed the two skulls shown, at the Lima Museum (Peru). The official explanation is that by binding a newborn baby's skull, the still-soft skull can be manipulated. But does this process also make it possible to double the standard cranial capacity? Or were the skulls already bigger than ours to start with? The eye cavities, which are not affected by binding, are also much bigger than those of modern Man.

long skulls described must belong to modern/Cro-Magnon Man. However, some of the long skulls display indisputable Neanderthal characteristics, in particular a receding forehead.

Where do the first Americans come from? Genetic studies have shown that although America was largely settled from Asia, some DNA data suggests that the roots of today's indigenous Indian people also lie in Europe – in Italy, Finland, Israel and Asia Minor (*Science*, Vol. 280, 24 Apr 1998, p. 520).

A Lame Example

Like the human family tree, the allegedly 55 million year old family tree of the modern horse which was invented at the end of the 19th Century (Marsh 1870) is used today as a prime example of evolution. A constantly increasing overall size, starting from an original 30 cm up to the present day horse is demonstrated. Parallel to this increasing size, the horse is said to have undergone a phased reduction in the number of toes.

After a further century of research, we know today that this evolution was not linear, but many species emerged who had even taken certain backward "evolutionary steps" and that various species which must in evolutionary terms lived subsequently to each other were, in fact, contemporaneous. Stephen Jay Gould confirms that all significant lines of descent for the odd-toed ungulates (that is the larger group of mammals to which horses belong) are the pitiful vestiges of earlier, more exuberant successes (Gould, 1998,

p. 97). And Gould stresses: (ibid p. 92) that any objective observer (must) recognise the decline as the most important feature of the horse revolution in the last 10 million years, in precisely that period in which according to the standard ladder model (with increasing complexity) perfection and a clear trend in the direction of a toe with a single hoof and shortened toes were achieved.

So where is the real cradle of the Horse? It is remarkable that it is in North America, where there was no longer a single horse by the time of Columbus's discovery, that the horse's prehistory is best preserved. The original horse, Eohippus, (Hyracotherium) was already discovered in Suffolk, England in 1838 and complete skeletons were found only in 1867 in North America (Utah, Wyoming). The first horse therefore existed on both sides of the Atlantic.

It is however strange that there has allegedly been no land bridge between America and Europe for at least 65 million years. But how then can Hyracotherium have lived on both sides of the Atlantic? In his book *Erdzeitalter* (Geologioc Era), Edgar Dacqué wrote that "the equine (horse) type has his origin in what was then a warm Arctic region, closer to America than Europe. This is why more and more specialised species emerged first there (America) and somewhat later here (Europe). Gradually the tribe spread out through Asia and to India..." (Dacqué, 1930, p. 515). Dacqué indicates that the original home of the horse was in today's Arctic regions between and/or north of America and Europe. Horses seem to have the same northern homeland as do our ancestors (Neanderthal, Cro–Magnon Man).

It is not only the first horse but practically all species of horse which can be proven to have existed in both America and Europe:

• At the beginning of the Miocene (20 million years) – Anchitherium.
• At the beginning of the Pliocene (5 million years) – Hypohippus.
• During the Pliocene (2 million years) Hipparion.
• During the Pleistocene up to 10,000 years ago – the modern horse.

Conventional teaching says that the modern horse (Equus) developed into his present form in America and migrated to the Old World 3 million years ago (Science, Vol. 307, 18 Mar 2005, pp. 1728–1730), across the Bering Strait through Siberia and into Mongolia – in the opposite direction to that taken by human settlement – a hardly convincing proposition. Horse died out in North and South America 10,000 years ago but was able to survive in Eurasia in the regions around the Black Sea.

Even if one assumes that there were in fact two enormous migrationary movements of horses from America to Asia, different development stages of the horse's evolution are still visible at the same time on both sides of the Atlantic (cf. photo 67). It is theoretically possible that horses spread out from the Bering Strait but they would have got no further than east Siberia and Mongolia, because otherwise the early horse species would also have spread to India and Africa. No, these horses quite simply lived in Beringia (the region of the dry Bering Strait), North America, Canada and Greenland, i.e. in what was once a single, warm, Arctic region which stretched as far as Europe. At this time, large areas of today's coasts of north America, Greenland, Europe (shelf regions, Greenland Bridge) and Siberia (e.g. Barents Sea) which are now 130 metres and more below sea level, were dry.

The sinking of Iceland by about 2,000 metres in earth's recent past at the end of the Bronze Age went hand in hand with the submergence and thus destruction of the Greenland Bridge. In *The Columbus-Mistake* I describe in more detail the scenarios which explain the flooding of the north American continent, as well as northern and western Europe.

In Europe, smaller horses, better adapted to the cold were used in the Young Stone Age to the Bronze Age, as shown by early depictions (incl. the Bayeux Tapestry c. 1200). Dramatic events must have taken place in the North Sea region because this former steppe, including the high Dogger Bank, was flooded by intense storm waves with constantly rising levels of water. "Wild horses, as they have been painted by humans on the cave walls at Niaux and Lascaux, moved across the North Sea steppe to west Norway and had to remain there when the sea returned" (Fester, 1973, p. 32). In fact, these were short, resistant horses (ponies) with great stamina. Naturally speaking, these horses have no connection with the high mountain valleys f the Scandinavian fjords. There were isolated for centuries because of the flooding of the North Sea savanna and are therefore regarded as an individual race of horse. The Vikings brought these animals to Iceland, so that the animal is also known as the "Icelandic Horse".

The various co–existing horse species were destroyed by catastrophes and like many other animals and humans, washed into flood deposits. The Icelandic horse and the modern horse escaped the cataclysm because humans nurtured and cared for the remaining populations.

It is not only the spread of horse species on both sides of the Atlantic which bears witness to the existence of the Greenland Bridge, but also the development of large mammals which supposedly only took place after the dinosaurs became extinct.

A rich, highly differentiated mammal fauna was discovered in 1878 in Cernay near Rheims (Lemoine in *Soc. d'Hist. Nat. de Reims*, May 1878) and "soon afterwards corresponding fauna were found in the Puerco layers of New Mexico (in America, HJZ). Later finds in Transylvania, Swabia, Switzerland, England, Utah and Wyoming showed their wide incidence. Ten genii are common to Europe and America..." Johannes Walther, Professor for Geology and Palaeontology at the University of Halle writes: "One could believe that the Eocene (55–36 million years) mammal fauna of the Cuvier catastrophes were separated by the chasm of time of the Calciferous" (Walther, 1908, p. 481).

In other words, there was a uniform development of highly evolved mammals during the Tertiary on both sides of the Atlantic. So this happened at a time when the continents had supposedly already been widely separated from each other for many millions of years. Without a land connection (Greenland Bridge) identical mammals on two different continents separated by a wide ocean is unthinkable – or did all of these animals migrate through Siberia? But during the "Great Ice Age", an alleged million years ago, when *Homo erectus* was supposedly starting to conquer Europe, there was already an animal in Eurasia's then savanna, which was previously only known in America: the puma. To the surprise of the scientists, this big cat was excavated in the Werra Valley close to Untermassfeld in Thuringia, Germany, (Kahlke, 1997/2001), in the middle of an "Ice Age" mass grave containing animals of all different kinds such as giant hamster, horse, elephant, bison, cheetah, hyena, hippopotamus and jaguar. "There must have been a deluge (= superflood, HJZ) which gobbled up these countless animals" (German *Geo Magazine*, 07/2005, p. 126).

Now there is a scientific debate as to whether the puma did not come originally from Eurasia because another modern animal which today lives only in temperate regions of south and Central America beside the big rivers, also existed in Europe during the "Great Ice Age", namely the jaguar. This animal has occasionally been spotted in the southwest of north America, although in 1994 I saw a jaguar far further north, around 60 miles south of Kansas City, crossing Route 69 in front of me. Did these animals once live in what are today Arctic but were then savanna-type landscapes in a single Eurasian–American region and were these heat-loving animals simply displaced southwards because of the cooling climate, surviving in central and south America, whilst in Eurasia their path south was blocked by seas, lakes and mountains running in an east–west direction?

The nomadic saga of the Cheyenne (Müller, 1970) says that they originally lived in a land to the far north, in a place without ice or cold and, alongside them, there were two other types of people, one with hair all over their bodies and the other with white skins, with hair only on their heads, their faces and their legs. The hairy people went south first and then the people with the long beards left (nobody knows to where) and finally the red man made his way south too.

In his book Beliefs and Thoughts of the Sioux (*Glauben und Denken der Sioux*, 1970), Werner Müller points out that the Sioux tribes must have come from what is now an Arctic landscape in the far north and he underlines that the Delaware Indians also recall that a flood and earthquake catastrophe forced their forefathers to leave their old land, the north land or turtle land (cf. Fig. 33, p. 231). It had become cold and snowy there.

Theoretically, humans might also have been discovered in the "Ice Age" site in the Werra Valley near Untermassfeld alongside heat–loving animals, but for Kahlke the likelihood of this is zero. If such a find had been made, however, "all hell would break out" he prophesied, saying "and we would have to play a part in the vanity-filled hominid show" (*Geo* magazine, 7/2005, p. 152).

Primeval Family 5,000 Years Ago

The question of the origins of our ancestors and/or Stone Age Man is made more difficult because of falsifications and invented age datings of skulls and bone fragments. Parts of skulls led not only to reconstructions of entire individuals but resulted in the conjuring up of thousands of fictitious generations out of thin air. If one disregards the falsifications of the anthropologists, there are in fact a few older skulls which might date back to the time before the Deluge. However, the available data is too sparse to make any definite statements. But neither the Neanderthal nor *Homo erectus* nor Homo heidelbergensis belong to the pre-Deluge period and therefore do not represent stages of human evolution. In chronological terms, these supposed Stone Age people should be beside each other rather than lined up one after the other. Neanderthals and early-modern Man lived in the Young Stone Age and their presence in certain regions is the result of the prevailing climate. With the shifting of the climate zones or a drastic climate change, these people adapted to the new situation or migrated to other regions to avoid natural catastrophes. The sudden emergence of new types of human is therefore not due to an evolutionary process of macroevolution, and finds of human relics below metre–thick deposits are proof of catastrophic turmoil, meteorite impacts and tectonic shifts beneath the earth's crust. These processes caused the face of the earth to be completely transformed, sometimes in a matter of mere hours, at different times and with regional differences, so that a sort of ragbag of geological evidence – as geological maps do indeed look – was created.

It is likely that for a short period of modern Man's existence, all humans were genetically far more similar than was previously thought (*Science*, Vol. 294, 23 Nov 2001, pp. 1719–1723). If humanity were old, there should be evidence of greater genetic differences.

Did the first fathers of modern Man appear only 3,000 years ago? In order to develop a family tree in which all people living today can be seen to come from the same root, we need only 33 generations, if we take account of an average reproduction period of 25 years. Therefore we can account only 825 years for all 33 generations. If, when making these statistical calculations one also factors in geography, history and migration, our identical ancestor lived, at most, 5,000 years or 169 generations ago. This means that a single person 5,000 years ago was either the ancestor of all people alive today or his genetic line died out completely. He cannot have been the ancestor of only a few people living today. A research team used a computer to produce several different scenarios, factoring in differences in population growth, isolation of individual groups and local and mass migration. The results of this study show that our youngest common ancestor probably lived around 3,000 years ago (*Nature*, Vol. 431, 30 Sep 2004, pp. 562–566).

If one considers the galloping development of the world's human population and accepts a catastrophe horizon of only a few thousand years, the findings of an in–depth genetic study fit into this argument: humanity was almost completely wiped out at least once in the last few million years. This means that Man's ancestors must have lost a great deal of their genetic diversity, probably because the number of people was significantly reduced (*PNAS*, 1999, Vol. 96, pp. 5077–5082).

This finding contradicts a continuous human evolution. How is it possible that the theory of the evolution of Man which is so obviously wrong was able to develop at all? Only because of utterly brazen scientific falsifications!

6   Falsified Proof of Evolution

The circles and study groups of scientific advocates of evolution are as blind and fanatical as the representatives of religious sects when it comes to defending their pet theory. It is a fact that, in a great many cases, scientists have served up the most monstrous and incredible falsifications to the public with the tacit approval of their colleagues. The most astounding phenomena include numerous exhibits throughout the world which supposedly represent and prove Man's descent from apes. These proofs, however, are purely fictitious and stem from the fantasy of their creators. With pen and brush in hand, the evolutionists produce fantasy creatures. The fact that these pictures don't correspond with any fossils is a real problem for them but despite this, these fantasy creatures are presented as if they were real. Fossils which can't be found are quite simply manufactured to fit the theory. Science's reputation is exploited to perpetuate falsifications and deceptions in the realm of earth history and human science.

How a Pig's Tooth Became a Human

The breakthrough of evolution theory was assisted by a court case which took place in Dayton, Tennessee in 1925. The elementary school teacher, John Scopes, was facing prosecution because he had taught his pupils about human evolution. Many scientific heavyweights supported the charged evolution advocate, led by Prof. H. Newman from the *University of Chicago*. As evidence of evolution theory, this scholar presented one of the time's prime examples: Nebraska Man. This race of humans was said to have lived in Nebraska a million years ago.

But wherein did the scientific proof of the existence of Nebraska Man lie? In 1922, a man called Harold Cook discovered the fossil remains of this primeval human – nothing more than a single tooth. This was examined by the most famous scientists and estimated to be at least one million years old. Afterwards a huge quantity of literature was produced about this ancient race. The *Illustrated London News* sent a reporter to America to find out everything about this newly–discovered human race, after which the paper published an article with pictures of reconstructed Nebraska people. The entire body had been reconstructed on the basis of a single tooth (Fig. 45).

Hot scientific debate ensued during which some scientists ascribed the tooth to Pithecanthropus erectus (*Homo erectus*) while others asserted that it was more similar to the teeth of modern humans. In 1927, however, another part of the skeleton was found and it became clear that the tooth did not come from Nebraska Man but actually belonged to Prosthennops, an extinct species of American wild pig. The

Fig. 45: Fraud. A complete reconstruction of Man's predecessors in an American nature scene with camels in the background - created on the basis of a single pig tooth. Countless children now believe that this fantastical illusion of ape-like people from Nebraska is the truth.

palaeontologist, Prof. Dr. William K. Gregory admitted that a fundamental error had been made in an article in *Science* (Vol. 66, 16 Dec 1927, pp. 579–581).

Thereupon all depictions of Nebraska Man and his family were removed with the greatest possible haste from evolutionary literature and the museums. But by being presented at the sensational court case, the pig's tooth helped the breakthrough and victory in America of evolution theory in the eyes of the public, as documented in the film Inherit the Wind.

Gone With The Wind

Since the 1920s, fossils have been excavated in Chou Kou Tien, about 40 km south of the Chinese capital, Beijing. Between 1929 and 1937 a total of 14 skull fragments, 11 lower jaws, many teeth, a few skeleton bones and a large quantity of stone implements were excavated. The relics were estimated to be 300,000–500,000 years old. This pre-human was given the name *Homo erectus pekinensis* (formerly *Sinanthropus pekinensis*), *Peking Man* for short, who is still a firm link in the chain of human evolution.

Most studies of these fossils were carried out by Davidson Black until his death in 1934. The German scientist, Franz Weidenreich succeeded him and studied the fossils until he left China in 1941. The original fossils disappeared in 1941, when they were supposed to have been sent by ship to the USA. Is it a coincidence that the originals are no longer available, especially since there are not even any photographs of the originals?

A detailed compilation of all the facts relating to *Peking Man* was produced by the missionary and scholar, Patrick O'Connell who returned from China in 1947. He explained that the find site of *Peking Man* was a lime pit. This was why an early lime kiln was used here. The locals had the habit of killing apes and then eating their brains. When the hill somehow collapsed at some stage, people were buried underneath it, whose skeletons fossilised in the lime layers over the course of time. Ultimately, *Peking Man* was reconstructed with artistic licence, from a mixture of ape and human bones.

It is assumed that a Dr. Pei, who continued with the excavations alone during the Japanese occupation, had good reasons to make the fossils disappear for the models produced from them showed little resemblance to the sober descriptions of the skull by Marcellin Boul, Teilhard de Chardin and Abbé Breuil, who had appraised the finds independently of one another. Furthermore the fact that industry had been practiced in Chou Kou Tien on a wide scale in prehistory was deliberately kept from public knowledge (Criswell, 1976, p. 92).

It appears possible that the entire hill slope began to slide with the extraction of the lime. The lime kilns which were built a very long time before and heated with straw and reeds, was buried beneath the rubble. This scenario would explain the layers of ash which would otherwise be used to claim that *Peking Man* already knew how to use fire. The collapsing rock mass left hollow spaces beneath it and there were indeed bones found in such hollow spaces. This is where the idea that *Peking Man* was a cave dweller came from. It has also remained largely unknown that in the Chou-Kou-Tien lime quarry, apart from animal bones, the remains of a real human were found, probably killed by the mountain slide. The actual proofs of *Peking Man* bear the strongest resemblance to the remains of a large, now extinct baboon, which was once very prevalent in the area. They were hunted by man and their skulls were opened with force to remove the nourishing contents. Thus arose the horror story of our cannibalistic ancestors. But *Peking Men* were just normal Chinese (Criswell, 1976, p. 92f.). Seen in this light, a new find makes sense. In 1966 two further skull fragments were discovered, which fitted together with the other two fragments (found in 1934 and 1936). These parts were found at a higher level and seemed more modern than the other skull caps (Jia and Huang, 1990).

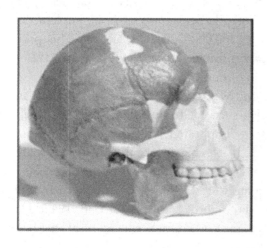

*Fig. 46: **Peking Man**. Human skull bone (dark) was combined with the jaw of an ape (light areas) to create an ape-like appearance.*

The Java Ape

Like *Peking Man*, *Java Man* is also supposed to have belonged to the *Homo erectus* (formerly *Pithecanthropus erectus*) species. The new term signifies that the "human" found on Java went upright, while the old name means "upright great ape". In 1891, the Dutch military surgeon, Eugene Dubois who was stationed in Java near Trinil, discovered a flat, very thick skullcap and three molars in a cave on a river bank.

It is currently in question as to whether the teeth are in fact from an orang-utan, i.e. a great ape. Java Man's age is uncertain but it is believed that the find is around 700,000 years old. The cranial volume is only about 940 cubic centimetres

In 1926, Prof. C.E.J. Heberlein from the Dutch state medical service said that he had found a new Pithecanthropus skullcap in Trinil in Central Java. This find, which was scientifically verified, proved to be the kneecap of an extinct species of elephant (Königswald, 1961).

From the outset, scientists disagreed about the finds. Some said that the bones came from a baboon or gibbon; others believed they were the bones of a great ape while another group claimed that they were human relics. Prof. Virchow from Berlin found that "there is no evidence that the bones come from the same creature".

The historian H.G. Wells then conceded that they were only ape bones and Eugene Dubois, who had trumpeted his find as the missing link between ape and Man, confirmed in 1932 that Pithecanthropus was not a human "but rather a gigantic, extinct species, similar to the gibbon") Dubois, 1937, p. 4; cf. Gould in *Natural History*, April 1990, pp. 12–24).

A very similar and more complete skullcap to the one found in 1891 was discovered by Gustav Heinrich Ralph Königswald near Sangiran in Java in 1937, and it was therefore named *Sangiran 2* or *Pithecanthropus II*. The specimen however has a cranial volume of only 815 cubic centimetre and is therefore even more apelike. Nevertheless, this find is regarded as significant. Overall this is a prime example of palaeo-anthropological swindling.

It is shocking that some finds can be deliberately wrongly interpreted, merely so that a success can be reported. The false reports however take root in the minds of interested laymen. Decades have usually elapsed before a retraction is made and it is then only read by experts.

It is also doubtful, indeed, it should be strongly contested that the reconstruction of the size of a brain and the shape of the entire head can be created on the basis of a skullcap from which the base of the skull is missing. Using a skullcap, anthropologists reconstructed not only the head's appearance but also the entire body.

As a result, not only has a fictitious member of some species or another been created, but millions of generations also rise up out of the depths of the past! There's method in their madness…

One Lower Jaw, Countless Generations

One may marvel at the so–called Heidelberg Man (Homo heidelbergensis) in reconstructed form in museums and there are pictures of him in most of the literature. In reality, however, no skeleton or even parts of one were found; in 1907 an extremely robust lower jaw with all of its teeth and a receding chin was discovered in the Mauer sand pit close to Heidelberg. The conditions under which the find was made were anything but ideal. If an anomalously old modern skull had been found under the same conditions, one would expect the find to be met with merciless criticism. However, since this jaw was deemed to fit into human evolutionary history, it was propagated without a breath of contradiction, as "proof" of the existence of many generations of human ancestors going back more than 400,000 years.

On the basis of his alleged age of 500,000–400,000 years, the find was first identified as *Homo erectus* and then as Homo heidelbergensis of the Homo genus (Schoetensack, 1908). When, after genetic studies, the Neanderthal was dropped as modern Man's forerunner, Heidelberg Man was upgraded to become the common ancestor of both Neanderthals and modern Man.

In other words, on the basis of this lower jaw, a distinct human race was presumed, which was then worked in as a connecting link in the chain of human evolution. For this, Heidelberg Man also had to be placed in time correctly. Initially his age was estimated to be 700,000 years, while today it is said to be 500,000–400,000 years as this fits better into the chain of evolution. Since nothing can be measured it is common practice to make fossil bones millennia older or younger at will.

All that one needs to create a human race from a lower jaw is a lot of plaster and a great deal of fantasy – and there in front of you from the depths of prehistory – is a plump, apelike creature whom one can now marvel at in museums as evidence of human evolution. No doubt about it!

But what is the true significance of the discovery of a particularly large, robust lower jaw with molars, which is of the same size as that of some modern humans (Wendt, 1972, p. 1622)? Aren't there people today with the same massive lower jaws? Over time, very few finds have been added to the Heidelberg jaw, and they all display a mixture of primitive anatomical characteristics (heavy brow ridge, thick skullcap) and modern characteristics (form of forehead and nasal bone). The group which is categorised in specialist literature as Homo heidelbergensis is really the same as the archaic *Homo sapiens*.

A famous scientist proved that an Eskimo skull had the same peculiarities and the same appearance as the lower jaw found. Another said that he had found an entire race of South Sea Islanders in the South Pacific who all had the same massive jawbones as Homo heidelbergensis (Criswell, 1976, p. 95).

In a grave mound near Toledo, Ohio, 20 skeletons in sitting position and facing east were found whose jaws with teeth were twice the size of those of modern humans

(*Chicago Report*, 24 Oct 1985). In another mound in Brush Creek Township, Ohio, the local historical society discovered eight skeletons, of which the smallest was 2.4 metres high and the largest 3.05 metres (*Scientific American*, 14 Aug 1880, p. 106). There have been other similar finds in North America.

On the 1569 Mercator map of the world giants are shown to live in southern Patagonia, Argentina. The chronicler, Antonio Pigafetta also reported giants in Patagonia whilst accompanying Ferdinand Magellan's voyage in 1520. Having brought two of these giants on board, it was noted that the tallest European sailor reached only to their waists (cf. photo 46).

These giants must surely have had jaws which were no smaller than those of Heidelberg Man.

The Long Term Deception

In 1912, the amateur palaeontologist, Charles Dawson presented his claim that he had found a cranial fragment, a jawbone and two molars at the edge of Piltdown near Brighton in the southern English County of Essex in a gravel pit. Although the jawbone more closely resembled that of an ape, the teeth and the skull were obviously human. A specimen was reconstructed on the basis of these few parts and was given the name *Piltdown Man*. The anthropologists were delighted and said the finds were 500,000 years old since, after all, this was the long–sought missing link between Man and ape...

This early human was reconstructed with great artistry, paintings were made and interpretations written. The bone fragments were exhibited in museums worldwide as the absolute proof of human evolution. Over a period of more than 40 years, numerous scientific papers were produced (*Science*, Vol. 40, 31 July 1914, pp. 158–60) and no fewer than 500 dissertations were written on the subject.

For almost 50 years human evolution was presented to the public as a fait accompli in museums, scientific publications and in the media. Children pressed their noses against the showcases and in turn told their own children of the miracle of evolution. The minds of several generations were befuddled as if with a narcotic, by this supposedly certain knowledge. Two entire generations of people in the 20th century staggered out of the fog of ignorance, intoxicated by new proof and findings, into Science's brave new world, illuminated by knowledge.

There were, however, individual scientists such as the zoologist Gerrit S. Miller (*Smithsonian Misc. Collection*, Vol. 65, 1915, p. 19) from the Smithsonian Institution in Washington D.C., who in 1915 described the jawbone as that of an ape. David Watson was of a similar opinion (*Nature*, Vol. 92, 1913, p. 319). But the wave of euphoria at this time over evolution theory which previously had been regarded with pity, washed away all objections because, after all, the proof was right there.

Only 37 years later was the deception uncovered. In 1949 a test with the newly–developed fluorine method revealed that the jawbone contained no fluorine whatsoever while the skull contained only minute quantities: this indicated that they had been in the ground for only a few years. Conclusion: the skull is not millions of years old but only a few hundred, at most (*Science*, Vol. 119, 26 Feb 1954, pp. 265–269).

The jawbone belonged to an ape which died at the beginning of the 20th Century. The teeth were posthumously implanted in the jaw. They had also been filed to more closely resemble human teeth. Then the bones were treated with potash and iron to make them into fossils. Treatment with acid removed this colouring. The team which discovered the fraud rightly asked: The evidences of artificial abrasion immediately sprang to the eye. Indeed so obvious did they seem it may well be asked – how was it that they had escaped notice before (*New Scientist*, 5 Apr 1979, p. 44)?

It was possible because this was not simply a mistake but a broadly staged swindle by palaeoanthropology. In spite of obvious falsification this find was intensively pushed by the media and published in lexica for 40 years as an alleged proof because the unproven ideology of evolution had no other means by which to establish itself. In other words, pseudo–scientific evolution used this propaganda to achieve its aims because the principles of macroevolution postulated by Darwin completely defeat common sense – because you can't make a Man out of an ape.

Everything about *Piltdown Man* which had been exhibited for 40 years was secretly removed with all speed from the showcases of the British Museum in London. But there is a further falsification which is still used in the biology classes of our secondary schools as proof of evolution and still comes up as an exam question.

Haeckel Already "Protsched"

What was once known as "recapitulation theory" has long since been written out of scientific literature; nonetheless, this theory is still presented as a scientific reality in some evolutionist publications. This theory propounded by Ernst Haeckel in the 19th Century said that living embryos go through (recapitulate) the evolutionary process which their pseudo-ancestors are said to have undergone.

Haeckel posited that during its development in the mother's womb, the human embryo displays the characteristic features of fish, reptile, mammal and finally human. Since then, it has been proved over the years that this theory was utter nonsense.

It is now known that the "gills" which appear during an early embryonic stage are actually the first phase of development of the middle ear channel, the parathyroid and the thymus. The embryonic part which was said to be an "yolk-sac" proved to be a sac which created blood for the embryo. The part which Haeckel and his supporters identified as a "tail" is, in fact, the spine which only during its formation resembles a tail before the legs

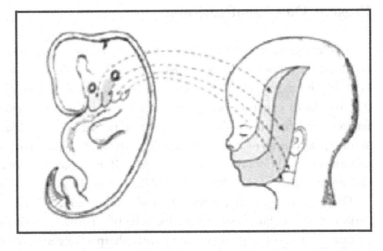

Fig. 47: Jaw Error. *Since in the propagated evolutionary series amphibians have gills but no pharyngeal arch, it is not clear why, according to evolution theory, the mammals who came later should once again have pharyngeal (gill) arches. According to Haeckel, the folds on the human foetus were wrongly interpreted as "gill slits" inherited from a distant ancestor. Human embryos definitely don't conform to the process laid down by "human evolution theory". Picture: Harun Yahya.*

take shape. These are universally accepted scientific facts.

George Gaylord Simpson, one of the founders of Neo--Darwinism writes: "Haeckel misstated the evolutionary principle involved. It is now firmly established that ontogeny (the development of the individual from fertilised egg to reproductive maturity) does not repeat phylogeny" (Simpson/Beck, 1965, p. 241). In an article published in the magazine *American Scientist* (Vol. 76, May/June 1988, p. 273).

Keith S. Thompson said that: surely the biogenetic law is as dead as a doornail. It was finally excised from biology textbooks in the fifties. As a topic of serious theoretical inquiry, it was extinct in the twenties.

Ernst Haeckel himself was good observer of nature but as eager for fame as Professor Reiner Protsch. This is why he faked his drawings to support the theory he advocated. Haeckel's fake drawings deceitfully illustrated the human embryo looking like a fish. When this was revealed, his sole defence before the American Senate was that other evolutionists had committed similar shameful acts:

After this compromising confession of forgery I should be obliged to consider myself condemned and annihilated if I had not the consolation of seeing side by side with me in the prisoner's dock hundreds of fellow--culprits, among them many of the most trusted observers and most esteemed biologists. The great majority of all the diagrams in the best biological textbooks, treatises and journals would incur in the same degree the charge of forgery, for all of them are inexact, and are more or less doctored, schematised and constructed (Hitching, 1982, p. 204). Faking everywhere you look! Since enough people had burned their fingers with the missing link, another myth was launched: birds were said to have evolved from dinosaurs.

Present-Day Falsifications

In October 1999 a sensational find was introduced and celebrated with a press conference, followed by colourful pictures in the November issue of National Geographic magazine: the missing link between dinosaur and bird. In *Science* it was said that thanks to exquisite preservation of soft parts, it's possible to see dragonfly wings, feathers, and fur on fossils estimated to be perhaps 120 million to 130 million years old allows palaeontologists to explore how feathers and flight evolved, as well as the relationship of dinosaurs to primitive birds. Both topics feature major scientific questions about the history of life. (*Science*, Vol. 279, 13 Mar 1998, pp. 1626–1627).

The long–sought and, since Darwin's time, sadly missed Missing Link (one of an endless chain required) was given the scientific name *Archaeoraptor liaoningensis*. It was an animal the size of a turkey, armed with sharp claws and teeth. As the first feathered dinosaur capable of flight it was believed to be the missing link between reptiles and birds.

Philip Currie of the *Canadian Royal Tyrrel Museum* let slip that although Archaeoraptor displayed all the characteristics required for flight, it was not known in how far these skills were developed. The pectoral girdle and breastbone resemble those of modern birds and its hands had already partially evolved to form part of a wing—like body part. Hollow bones, feathers and tail – possibly important for stability in the air – were also in evidence. However, the tail was quite long and stiff which would surely inhibit flight ability.

Archaeoraptor liaoningensis joined a group of feathered dinosaurs found in China in recent years. Exhibitions were organised, for example, in Washington and the missing links was presented to a marvelling public as proof of macroevolution amidst the flashbulbs of the photographers. The worldwide dissemination of this proof by the magazine National Geographic which is sold all over the world contributed to this manifestation of the new find of the century. We knew it all along and we were right: there is such a thing as macroevolution, the first proof had finally been found and it can now be presented…

For two years, this first find of a missing link was the focus of an enormous advertising campaign started by scientists and the media. Computer animations were shown on television, to demonstrate how a flightless animal could suddenly become a master of the air, as if a chicken had turned into an eagle. Had I made a mistake when I argued against the posited evolution of dinosaurs into birds in my by book *Darwin's Mistake* (2003, p. 190):

»Embryonic studies are claimed to suggest what reduces first in the progressing development from a hand into a wing is the outer fingers. This corresponds with the appearance of a bird's hand. It, nevertheless, is a fact that their alleged ancestors, the theropod, did not have outer fingers, the fourth and fifth finger of the hand. It seems, this rule has a decisive exception. Biologists now try to prove, carpal bone and finger of non-bird-theropod are equivalent to those of later birds in terms of form and joint. Mind you, both cases refer to absolutely different fingers! All interpretations of parallel development therefore must seem arbitrary. But biologists are desperate seeking for the missing link«.

Then two years later, in December 2000 a short article appeared in *Science* (Vol. 290, 22 Dec 2000, p. 2224): Early this year the flying dinosaur fell to Earth in a jumble of parts from two distinct fossils – a primitive bird and a dinosaur. It hurt us tremendously, says Jim Kirkland of the Utah Geological Survey, who worries that the public will think all feathered dinosaurs are faked.

Then on 29 March 2001 an exact analysis appeared in *Nature* (Vol. 410, pp. 539–540) explaining how the (artistically excellent) Archaeoraptor ("old predator") was faked.

And on foot of this report, the objections I had already raised in 1998 against the posited bird evolution were scientifically confirmed in 2002: "The PhD student Julie Nowicki of the University of North Carolina in Chapel Hill had opened up eight–day old ostrich eggs. Nowicki found that around this time, the bird embryos develop fingers. It may be clearly seen that the three fingers of the bird's hand develop in the same way as the human forefinger, middle and ring fingers. The three fingers of the dinosaur, however, correspond to thumb, forefinger and middle finger…" (*BdW*, 16. Aug 2002). This PhD thesis exactly confirms, three years later, my arguments in "Darwin's Mistake".

But there are also several other reasons why dinosaurs cannot be the forefathers of birds, one of which is that the bird's complicated lung cannot have developed from that of theropod dinosaurs (Zillmer, 203, p. 190). This "objection can currently be neither confirmed nor contradicted as the organ is not preserved in fossil form. There was however no other animal out of whose lung the extremely complicated organ of the birds (which is different to that of every other living animal group) could have evolved" (*SpW*, 4/1998, p. 43).

There are many indicatations that today's birds did not evolve from dinosaurs (*Nature*, Vol. 399, 17 Jun 1999, pp. 679–682).

Selection and Mutation

Gregor Mendel's (1822–1884) laws of heredity are today the basis for experimental genetics and evolution, that is to say, microevolution, for supposed macroevolution – the evolution beyond the level of the species (Mayr, 1991, p. 319), in other words the creation of new species – is always only "proven" by proxy on the basis of examples of microevolution (adaptation of species to environmental conditions).

A typical example is the famous Darwin finch. Darwin studied the various races of *Galápagos Finches* and came to the conclusion that these animals could transform into another species through a sequence of small deviations. How did he arrive at this completely false conclusion? The answer is simple: he did not yet know Mendel's laws of heredity. According to these laws the so–called "hidden characteristics" skip one or more generations to re–emerge later. When they re–appear they are unchanged and exactly as they were before and therefore do not represent a newly acquired characteristic. What Darwin thought was a new characteristic in finches was in fact simply a new combination of recessive characteristics which already existed in the ancestral line. The main objection to the theory of natural selection is that it cannot produce new characteristics. Selection chooses only those characteristics which are most suitable according to Mendel's laws, such as colour or camouflage.

Biologists before Darwin were already familiar with natural selection as a process of nature. It was defined as "a mechanism that keeps species unchanging without being uncorrupted". This observation is correct. What is wrong is Darwin's idea that this selection contains evolutionary power. The English palaeontologist, Colin Patterson, emphasises: "No one has ever produced a species by mechanisms of natural selection. No one has ever gotten near it and most of the current argument in neo–Darwinism is about this question" (in: *Cladistics*, BBC, 4 Mar 1982).

Let's take an example of natural selection: a herd of gazelles is threatened by lions. Naturally, those gazelles which can run faster survive. But this process will never transform the gazelles into another living species. Gazelles will always remain gazelles, irrespective of how long this selection process lasts, even if perhaps gazelles with particularly "fast" legs are the end result. Since natural selection is not a conscious process, there can be no deliberate selection to achieve greater complexity. Assertions to the contrary are crass attempts to cheat.

Thus natural selection does not lead to more complex systems but is only effective for change as microevolution. Through this mechanism the existing species is solely protected from degeneration. Moreover natural selection does not have the ability to transform one species into a different animal in the sense of macroevolution. That is the reason why Neo–Darwinism had to elevate mutation as the cause of beneficial changes to the same level and place it beside natural selection.

In order to provide positive proof of mutations going beyond the species, Darwinist geneticists have for decades been breeding fruit flies (Drosophila melanogaster), since these insects reproduce very rapidly and mutations can therefore be seen quickly. It was discovered that exposure to radiation considerably increased the number of mutations. The radiation caused fruit flies with red eyes to appear alongside those with black eyes; other had smaller, larger or more wings than usual. Many different specimens were created. In addition, generations of fruit flies were exposed to extreme conditions of heat, light, darkness and chemicals (Pitman, 1984, p. 70). All of these fruit flies were completely inbred so many different varieties appeared. However, despite countless attempts, these were still genetically the same fruit flies as at the beginning.

The geneticist, Gordon Taylor, wrote that though geneticists have been breeding fruit—flies for sixty years or more in labs all round the world. But they have never yet seen the emergence of a new species or even a new enzyme (Taylor, 1983, p. 48). In practice mutants die, are sterile or tend to revert to the wild type (Pitman, 1984, p. 70).

The same goes for human evolution. All mutations observed in humans have had harmful effects: physical deformity or lingering illnesses such as Down's Syndrome, albinism, dwarfism or cancer. This can hardly be an evolutionary mechanism. Nor did the Hiroshima Bomb produce any positive mutations, but rather sterility and deformities.

In her book *Evolution*, Ruth Moore (1970, p. 91) says that the work in many laboratories showed that most mutations are harmful and the most drastic are even commonly fatal. They are going in the wrong direction, so to speak, in the sense that every change in a harmonious well—adapted organism has a negative effect. Most carriers of far—reaching mutations do not remain alive long enough to pass on the changes to their offspring.

If one carries out an unsystematic change to a robot or a specialised organism, the mechanism will certainly not be improved but will most likely be damaged or, at best, show no effects. In the same, mutation does not produce positive effects.

Nevertheless, Evolutionists calling themselves Creationists (= theists: hybrid believers who believe in the Creation followed by evolution) also write that there are positive mutations, "which under certain circumstances increase the fitness of an organism … examples include the development of manifold resistances to antibiotics or pesticides and changes in bacteria which lead to the degradation of new substances" (Junker/Scherer, 2001, p. 102).

Most people can understand the example of apparently progressive resistance to antibiotics. Today it is recognised that the use of antibiotics leads to the building up of resistance in bacteria. It happens again and again that patients are infected by bacteria which are resistant to several antibiotics (multiresistant).

Therefore the preventive use in battery farming is controversial and is already rightly prohibited in some countries. But this has absolutely nothing to do with positive mutation or evolutionary development. An antibiotic is a metabolic product of microorganisms which kill bacteria (bactericidal effect) or impede their growth (bacteriostatic effect). The mechanism functions in such a way that most bacteria which come into contact with the antibiotic die, while the rest, on which the antibiotic has no effect, reproduce rapidly and soon make up the entire population, all of which is immune to antibiotics.

When one kills bacteria they are as dead as fish that go for a walk on dry land and want to become amphibians. However, the transmission of existing resistant genes is possible. This is where Evolutionists see an evolution of the bacteria through adaptation. This superficial interpretation is not the truth.

For the Israeli biophysicist and professor, Lee Spetner (2001) there are two mechanisms which have nothing to do with evolution:

• Transfer of resistance genes already extant in the bacteria.
• The building of resistance as a result of losing genetic data because of mutation.

Some microorganisms are endowed with genes that grant resistance to these antibiotics. This resistance can take the form of degrading the antibiotic molecule or of ejecting it from the cell. Organisms having these genes can transfer them to other bacteria making them resistant as well. Although the resistance mechanisms are specific to a particular antibiotic, most pathogenic bacteria have succeeded in accumulating several sets of genes granting them resistance to a variety of antibiotics. But this mechanism is no proof of evolution: The acquisition of antibiotic resistance in this manner is not the kind that can serve as a prototype for the mutations needed to account for Evolution. The genetic changes that could illustrate the theory must not only add information to the bacterium's genome, they must add new information to the bio-cosmos. The horizontal transfer of genes only spreads around genes that are already in some species (Spetner, 2001). So again, this is a case of ever-present and ongoing microevolution.

If antibiotics are already used during animal fattening, bacteria are being directly bred which are immune to one or more of the media being use. It is then no use for the people who eat this meat to take the corresponding antibiotic if it has been attacked by these cultivated bacteria. In other words: the more often and intensively antibiotics are used in animals and humans, the more ineffective they become. There should, therefore, be a general call for a ban on the use of these medications in animal farming. When experts speak of positive mutation and thus proven evolution with regard to these mechanisms, it is a (deliberate?) case of misinformation.

Three reasons may be presented to explain why mutation cannot be drafted in to support evolution theory (Yahya, 2002, p. 63):

• The direct effect of mutations is harmful. Since they occur randomly, they almost always damage the living organism that undergoes them. Reason tells us that unconscious intervention in a perfect and complex structure will not improve that structure, but will rather impair it. Indeed, no 'useful mutation' has ever been observed.

• Mutations add no new information to an organism's DNA. As a result of mutations, the particles making up the genetic information are either torn from their places, destroyed, or carried off to different places. Mutations cannot make a living thing acquire a new organ or a new trait. They only cause abnormalities like a leg sticking out of the back, or an ear from the abdomen.

• In order for a mutation to be transferred to the subsequent generation, it has to have taken place in the reproductive cells of the organism: A random change that occurs in a cell or organ of the body cannot be transferred to the next generation. For example, a human eye altered by the effects of radiation or by other causes will not be passed on to subsequent generations.

When, however, someone asserts that it is trifling mutations which have a positive effect on the development of species, it must be noted that each cell possesses an entire arsenal of control and repair mechanisms to keep the frequency of mutation as low as possible.

This is a necessary self–protective measure because otherwise the species would degenerate relatively fast, as minor, random damage to the DNA would lead to defects and, through heredity, to degeneration and infertility. Our bodies are therefore adapted to prevent mutation, or at least to keep it as low as possible and, where possible, repair the defect. Mutation is therefore not evolution's magic wand!

Evolutionists also invented the myth of "vestigial organs" which is repeated over and over again, like a mantra, in the literature. For decades we were expected to believe that the bodies of some creatures contain a range of organs with no function. These were inherited from ancestors and had gradually withered through lack of use – the garbage of evolution, so to speak. In 1895 the German anatomist R. Wiedersheim compiled a list of vestigial organs, which numbered almost 100 including the appendix.

But this is by no means useless and redundant but is a lymphatic organ which fights infection in the body. With growing scientific knowledge it was established that almost all of the organs on the list did, in fact, have important functions.

That is why this unscientific thesis of "vestigial organs" was buried at dead of night like other fairy tales of evolution theory. The "normal" person still believes today that we have a vestigial tail at the end of our spines. However, the coccyx, as a static support for the pelvis, is in fact indispensable. And Darwin was also wrong when he described the semi–lunar shaped fold of the eye as a vestigial organ because this fold serves to clean and lubricate the eyeball.

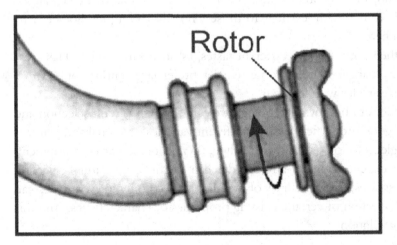

Fig. 48:
Engineering. This bacteria motor can perform up to 100,000 continuous rotations per minute. Such a motor, made up of 240 elements, can not have been created by innumerable mutations. ("Scientific American", September 1971).

Natural Selection

At the time Charles Darwin recognised the shortcomings of his evolutionary theory which were already under discussion and therefore propagated a further theory. He rightly saw that there are typical attributes which can only be explained by deliberate, i.e. consciously-steered, preference.

He called this principle "natural selection" and applied it in two areas. Firstly, Darwin went along with the prevailing conviction that the man's power of reason is superior to that of the woman and that male intelligence is greater and more outstanding than that of the female.

Secondly, Darwin tried to use the principle of natural selection to explain why since Man first appeared, he had no pelt, i.e. that he was born almost without hair. How could nakedness be an advantage?

Darwin explained the assumed superiority of the male's intellect over that of the female's by saying that males have always had to fight for females and because of this ongoing struggle the male developed intelligence superior to that of the female. Therefore the man is more intelligent than the woman. I would take the liberty of saying that if somebody today were to utter a sentiment even close to this, it is more likely that he would be locked up in a mental hospital than that such an idea could manifest itself as the basis of a scientific theory. But, according to Darwin, this fighting male acquired greater intelligence with all the fighting, which he would of course pass on – but he still had a pelt. And now natural selection comes into play because, allegedly, females preferred male great apes with less hair and as a consequence their children had less and less hair.

Charles Darwin's ancestors probably didn't fight long enough for a female to acquire a higher level of intelligence, otherwise he would have spotted this classic circular argument: on the one hand, the male's superiority is based on the fact that he had to win his women in battle and, on the other hand, the reason why humans aren't hairy is because women preferred men with less hair to those who still had a pelt. So – exactly who is choosing whom here?

Didn't the females then not also have different tastes, as do women today? Has there really been such a great change? Hardly! Some females prefer large anthropoids with a smooth pelt; others prefer males with less hair.

Applying Mendel's Laws to botany and zoology we can, through careful selection and breeding, bring out the good properties of many different varieties of animal and plant.

If, however, one neglects to look after the breeding, there is no upward development, but rather unavoidable degeneration, as any breeder can confirm. So natural selection does not produce new species; using the laws of heredity, at best (only) new varieties can be created through the selection of certain, existing properties or characteristics. And this is micro and not macroevolution!

Fossils vs. Evolution

According to evolution theory an existing species transforms into a new species over the course of time. Each of these transformations is supposed to take place gradually over millions of years. If this were the case, there should be countless transitional species for each of these countless transformations.

On the one hand, there must have been transitional forms in the shape of mongrel creatures in the past which, for example, displayed the characteristics of both fish and reptile and, on the other hand, there must be proof of millions of stages of development with regard to the organs and the extremities. But all of the known fossils and all animals living today appear to be fully developed and perfectly adapted.

If evolution theory is correct: so must the number of intermediate varieties, which have formerly existed, be truly enormous. Why then is not every geological formation and every stratum full of such intermediate links? Geology assuredly does not reveal any such finely–graduated organic chain; and this, perhaps, is the most obvious and serious objection which can be urged against this theory. The explanation lies, as I believe, in the extreme imperfection of the geological record (Darwin, 2000, p. 357f.)

At the time, Darwin hoped that sometime in the future some intermediate link would be found. However, the search which broke out and has continued up to the present to find intermediate links in fossils has not come up with a single such find, in spite of the millions of fossils which have been discovered since then. Darwin's doubts have therefore evolved, so to speak, into a certainty: macroevolution, unlike microevolution, cannot be proven from any fossils.

The British palaeontologist, Derek V. Ager, concedes the following fact: If we examine the fossil record in detail, whether at the level of orders or of species, we find — over and over again — not gradual evolution, but the sudden explosion of one group at the expense of another (*Proceedings of the British Geological Association*, Vol. 87, 1976, p. 133).

Intermediate links have not been found but rather the sudden appearance of new, fully–formed animals without any sort of intermediate link. Such a scenario can also be demonstrated in Cambrian layers (590–500 million years).

This wonderful phenomenon is described in geological literature as the "Cambrian explosion" because there is no evidence of organic life in the Precambrian (earth's earliest time) and then suddenly there is a swarm of life of all kinds. Life formed without any long-lasting period of evolution and without transformation of species – as proven by the "ideal" fossils without intermediate links found in geological formations.

Most of the life forms found in the Cambrian layers suddenly display complex organ systems such as eyes, gills and other highly developed structures, which are in no way different to those of their modern day counterparts. It is important to note that during the early Cambrian epochs, the existing animal genii were clearly distinct from one another as they are today, around 600 million years later. The diversity of species should really have developed like the branching of a tree from a single trunk but the individual animal trees appeared parallel to each other, just like rows of freshly planted seedlings without a common root.

It is as though they were just planted there, without any evolutionary history (Dawkins, 1986, p. 229).

In the journal *Science* (Vol. 293, 20 Jul 2001, p. 438f.) it is confirmed: The beginning of the Cambrian period saw the sudden appearance of almost all the main types of animals (phyla) that still dominate the biota today.

According to Charles Darwin the Cambrian explosion – with the sudden appearance of organs which seemed to have been planned by an engineer – should have rung the death knell of the evolution hypothesis. He himself was convinced: If it could be demonstrated that any complex organ existed, which could not possibly have been formed by numerous, successive, slight modifications, my theory would absolutely break down (Darwin, 1859, p. 206). This organ exists!

The trilobites appeared suddenly and possessed a complex eye, comprising hundreds of honeycomb–shaped individual eyes with a double lens system – an optimum design. This honeycomb structure of the trilobite eye has remained preserved over 600 million years up to the present day as insects such as bees and dragonflies have the same eye structure as trilobites (Gregory, 1995, p. 31).

When one further considers that many animal species, such as the salamander, have not changed since the time of the dinosaurs, or that spider's silk has apparently remained unaltered for 125 million years, one must ask oneself: what happened to evolutionary development?

In the Petrified Forest National Park in Arizona fossilised tree trunks tower even out of the slopes of today's table mountains – the former mud layers – 200 million years later – proof of superfloods in today's desertified region (cf. Photos 29–32). To the surprise of the experts isolated examples of fossilised bee and wasp nests were found in such fossilised trees (*The Arizona Republic*, 26 May 1995, p. B7).

However, bees and the plants they require allegedly only developed 140 million years after the uprooting of these trees.

Either the dating of the geological layers or evolution's time ladder is wrong. This is why such finds have not been scientifically publicised.

The finds, documented by experts, and the empirical chain of evidence reveals that evolution theory is a lie deliberately maintained and upheld by scientists of earth and human history.

The consequence of the described findings showing:

• Darwin was wrong!
• But also the geologic time scale is wrong!

Ideology, Racism and Terrorism

In his book *The Descent of Man*, Darwin daringly discusses the greater differences between the men of distinct races and placed Negroes and Australian Aborigine on the same level as gorillas. In this case, it is easy to assume that races developed differently.

There are, in fact, many cultures, but no races. This scientific fact is based on in–depth studies of molecular biology. Of course there are differences between people but these are far more marked between individuals than between peoples. Race is a real cultural, political and economic concept in society, but it is not a biological concept, and that unfortunately is what many people wrongfully consider to be the essence of race in humans – genetic differences in the words of Alan R. Templeton of *Washington University* in St. Louis (*SpW*, 9 Oct 1998).

In spite of differences in external appearance the genetic differences between the large human groups are so small and their characteristics so often overlap, that division into races is neither feasible nor sensible. Racism has no genetic foundation but is based on a psychological need for unassailable superiority and dominance. The differences between people have far less to do with genetic variety than with different mechanisms for transmitting technical and social innovations (Cavalli-Sforza, 1999).

That people across ethnic boundaries are genetically far more similar than was previously thought, is shown in the painstaking Chromosome 21 analysis of people from 24 different ethnic groups (*Science*, Vol. 294, p. 17). Irrespective of this fact, the existence of human races has been accepted as a reality since the 19th Century. Unfortunately the race issue is still being used today to create or demonstrate prejudices so that certain political goals and the associated egoistic desire for power may be achieved.

Adolf Hitler in particular used race in an inhuman way. A "master race" was to be given a privileged place on earth. In certain centres, people who bore the desired "Aryan" characteristics were selected, isolated and used for breeding. Adolf Hitler expressly cited Charles Darwin's hypotheses and saw evolution theory as a justification of his actions because in Hitler's view other "non–Aryan" races were subhuman's who, according to Darwin, were in any event doomed to extinction. Hitler therefore regarded himself as the executor of the postulated evolution mechanism. Millions of people were victim to this madness.

Supposing that living beings evolved in the struggle for life, Darwinism was even adapted to the social sciences, and turned into a conception that came to be called "Social Darwinism". Social Darwinism contends that existing human races are located at different rungs of the evolutionary ladder, that the European races were the most advanced of all, and that many other races still bear simian features (Yahya, 2002, p. 46).

Whilst the fascists are on the right-wing of Social Darwinism, the left-wing is occupied by the Communists. The Communists were always outspoken advocates of Darwin's theory and demonstrated this in their so-called dialectical materialism according to Karl

Marx and Friedrich Engels.

The reading public is sold the ideology of evolution as a guaranteed scientific truth. The racism embedded within it piggy–backs onto this theory and is projected into the outside world, to be seen by the reading public as supposedly real, scientifically proven, so to speak, even though the theory is only a mental Fata Morgana.

Pushing these anthropological wafts of mist aside, the competent Ashley Morgan writes that until recently, most anthropologists were still certain that race has some physical reality in nature. But the expression "race" as it is commonly used with reference to people, is scientifically untenable and that so, as it is commonly used, it has no relation to reality (Morgan, 1974; cited in Friedrich, 1994, p. 16).

The reason why "races" were felt to be a reality by our ancestors and by many people today, lies in the manifold differences between the peoples of different regions. Different characteristics such as skin colour, body size, the shape of the eyes, the body, the face and other details often lead us to guess, at first glance, where the person comes from. In each continent, many of these characteristics are homogeneous, thus creating the impression that there are "races". These differences are at least partially genetic. Skin colour and body shapes are the least heritable. They are almost all due to climatic differences (Cavalli-Sforza, 1999, p. 22).

Comprehensive new studies show that the classification of people into races and by ethnicity is useless (*Nature* Genetics, Vol. 36, pp. 54–60), in particular on the basis of skin colour. Only in 1958 was it discovered that melanin caused a change in the skin pigmentation cells of laboratory frogs. Special cells, the melanocytes, produce melanin, the skin's most important pigment. They are, however, not evenly distributed across the surface of the body.

Thus the soles of the feet have fewer such cells than do the upper sides of the feet. Together with the cornea, melanin represents the body's most important protection against UV–radiation. If the skin is exposed to sunlight or to UV-radiation in a solarium, the melanocytes increase the production of melanin. In other words: the production of a large quantity of melanin makes the skin dark; low production makes it light.

This has nothing to do with a gradual evolutionary adaptation to varying intensities of sunlight. Within isolated groups of individuals, certain characteristics which are already present with high frequency necessarily become more marked in view of selection factors (isolation, climate).
Strangely, Eskimos have "brown" skin similar to the Basques. Yellowish skin is not created by any additional pigmentation but is an effect caused by a thicker epidermis.

Thus Man's original colour is not yellow, white or black, but brown. Light–skinned people have less and dark–skinned people have more black pigment, melanin, than brown–skinned "prehistoric Man".
Albinos display a defect in that they are incapable of producing melanin. Human albinos are therefore white-skinned but in the animal kingdom a pure white animal is not an

albino but is "leucistic". On average every 10,000th baby born in nature is an albino. Partial albinism is more common in humans than in animals.

A different skin colour is not a biological "signifier of race". Nor does skin colour change over long periods of time through evolution by slow adaptation to the sun's rays.

As already mentioned, in only 13 generations of isolation, morphologically different salmon developed in a single lake (*Science*, Vol. 290, 2 Oct 2000, pp. 516–518).

With the racism generated by Darwinism, the basis of ideologies was established, which would throw our world in the 20th Century into the bloodiest conflicts it had ever seen: Nazism and Communism. In his book *The Dialectic of Nature*, Engels showed the indissoluble link between evolution theory and Communism. When we think of the Communist concept of "dialectical contradiction", we must also remember the 100 million people who fell victim to Communism.

Mao, who established the Communist order in China and was responsible for the deaths of millions of people, publicly stated that "Chinese Socialism is founded on Darwin and the theory of evolution". Darwinism is a root of violence which has brought only misfortune to the people of the 20th Century.

Nevertheless, just like these ideologies, Darwinism is defined by an "ethical understanding" and a method which can influence world views. The basic concept behind this understanding and this method is "if you're not with us, you're against us". It was the same attitude which led to masses of people being burned in the Middle Ages. This attitude may be explained as follows: there are different beliefs, different world views and different philosophies in the world. There are two ways in which they may be reconciled:

1. Cooperation: they can respect each other's existence, try to make contact and display a humane attitude.

2. Confrontation: they can decide to fight, argue and harm each other.

*Fig. 49: **White Skin and Hair.** A Nigerian albino beside a "normal" Nigerian female.*

The horror which we call terrorism is expressed in the second viewpoint. If we take Darwinism away, there is no philosophy of conflict.

We are supposed to believe that terrorism goes hand in hand with the religious ideas and symbols of particular regions (e.g. Islam). Those who even hint at such thoughts are really Social Darwinians and are using religion to cover up the attainment of economic and political interests and goals. For this reason the root of terrorism which is afflicting our world is not to be found in any of the monotheistic religions but in Darwinism and Materialism. The result is confrontation instead of cooperation.

The only reason why Darwin's theory, despite being obviously disproved by progressive scientists, is still being defended is because of its close links to Materialism. Early Capitalism gratefully adopted Darwin's idea, because it provided the seal of approval for remorseless maximisation of profits and merciless competition – whoever fell by the wayside was simply not "fit" or not strong enough and didn't belong to the "chosen". Darwinism was a fabulous excuse for anybody who was prepared to stick at nothing.

The philosophy of Materialism is one of the oldest concepts in human history. If someone is firstly a materialist and secondly a scientist, he will not deny Materialism but will try to uphold and preserve it, by defending evolution without any regard for the cost (Lewontin, 1997, p. 28). Prof. Reiner Protsch's actions perfectly illustrate this point.

Robert Shapiro, a professor of chemistry and DNA expert explains the belief of the Evolutionists and the underlying materialist dogma as follows: Another evolutionary principle is therefore needed to take us across the gap from mixtures of simple natural chemicals to the first effective replicator. This principle has not yet been described in detail or demonstrated, but it is anticipated, and given names such as chemical evolution and self–organization of matter. The existence of the principle is taken for granted in the philosophy of dialectical materialism (Shapiro, 1986, p. 207).

Evolutionist propaganda in the mass media is the consequence of this ideological demand. Since evolution is regarded as a necessity, those circles who set the the standards in science made it into a dogma, the contradiction of which is taboo. Members of the academic elite in the western world are forced to publish their writings in certain scientific journals, to achieve and retain their academic standing. All journals which deal with biology and/or evolution are controlled by evolutionists who prevent the appearance of any anti–evolutionary views in their publications.

The public at large is not aware of this systematic brainwashing and propaganda and considers evolution to be a scientific fact. Laypeople which for generations have been snowed under by corresponding reports naturally begin to believe that in spite of all the nebulous, fairy tale–explanations it provides, evolution is a fact. To deny evolution is proscribed in the scientific community and seen as displaying contempt for fundamental realities. Authors who formulate critical thoughts are ignorant and would have been burned at the stake in the Middle Ages. This is why, despite numerous defects and errors, particularly those which have been uncovered since the 1950s and conceded even by the Evolutionists themselves, there is barely a word of criticism spoken against evolution theory in scientific circles or in the media.

Thus, even in scientific papers the "Move from Water to Land" – one of the evolutionary phenomena with the least proof – is "explained" with laughable simplicity. But a fish cannot live for more than a few minutes outside water. If generations of fish tried visiting dry land, they would all have died within a few minutes, even if this process had repeated itself over millions of years. The reason is that complex organs, such as the fully developed lung in birds, cannot form suddenly out of the blue. On the other hand, a slowly developing lung in each of the required intermediate links, in keeping with evolution theory, would not be capable of functioning. There has never been such a thing as a partially or half–developed lung and no such specimen has ever been found in fossil form.

The fairy story of Darwin's egoism is also repeatedly brought up in the media. But it is not rivalry and merciless battle which are the successful strategies of evolution; it is communication and cooperation which have allowed us to progress. This must be clear since confrontations cause great friction loss and useful lifespan is shortened. This is why, for example, assembly belt work in factories was revolutionised. The "weak" no longer perform a monotonous, repetitive assembly line job, but accompany a work process where possible from start to finish so that, even as a weak link in the chain, they have a share in the success in the form of cooperation. Unlike with confrontation (exploitation of the weak) cooperation achieves better results and the individual is a happy part of the whole (cf. Ripota, 2002).

Here is an example of cooperation from the animal kingdom: at times when food is scarce, a certain type of amoeba (Dictyostelium discoideum) does not eat up his fellows until only the strongest remains. On the contrary, these amoebas begin an extremely cooperative course of action. Countless individuals climb on top of each other until a type of stalk is formed (slime–mould aggregation). Around 20% of the individuals forming this stem die, the rest turn into spores which are borne away by the wind. They have the chance of one day landing on fertile ground. The amoebas, who form the core and die, have sacrificed themselves, not selfish. A Darwinian would say that the stronger amoebas had climbed over the bodies of the weaker ones and survived. This example serves to illustrate the Darwinians' extremely cynical outlook.

With each phase of the arguments presented, evolution is seen to be not just a lie, but also a mental dead-end road of our social and economic development, which is of real benefit to few. Let us desist from this way of thinking and turn towards cooperation, not just in the western world but amongst all peoples thereby – and especially – showing tolerance for all the different religions. Then terrorism, or that which is so labelled (often falsely) by our governments, has no chance. As a result, we would have achieved real development as a result of tolerance (love of our neighbour) and cooperation.

"Darwinism is a teaching from the 19th Century which was jointly responsible for the dreadful events of the 20th Century. We must have the courage to divest ourselves of yesterday's convictions and re–think so that the 21st Century is better" (Ripota, 2002).

To this end it is necessary to recognise that evolution theory with the embedded idea of the development of humans and classification into races is a lie. This pseudo-scientific evolution serves as a religious surrogate to oppress and manipulate not just individuals but whole peoples, north and south, east and west.

Epilogue

The quiet after the storm is the calm before the hurricane. One of the Evolutionists most common arguments is: either one believes in evolution theory or one falls into the hands of religious maniacs. That is nonsense! But these slogans demonstrate how little many scientists really think of their explanations, making use of Inquisition–like methods in order to avoid controversial debate at any cost.

But Darwin's evolution theory is long a myth which only lives on today because people of all levels of education are told in big, bold pictures, with simple words and a scattering of jargon, about the origins of mankind.

Illustrative examples and new ways of thinking have revealed evolution theory to be a modern fairy tale. That the truth, buried under science's garbage is not recognised as such, is down to the pressure mechanisms which the American molecular biologist, Jonathan Wells, describes as follows:

Dogmatic Darwinists begin by imposing a narrow interpretation on the evidence and declaring it the only way to do science. Critics are then labelled unscientific; their articles are rejected by mainstream journals, whose editorial boards are dominated by the dogmatists; the critics are denied funding by government agencies, who send grant proposals to the dogmatists for "peer" review; and eventually the critics are hounded out of the scientific community altogether. In the process, evidence against the Darwinian view simply disappears, like witnesses against the Mob. Or the evidence is buried in specialized publications, where only a dedicated researcher can find it. Once critics have been silenced and counter–evidence has been buried, the dogmatists announce that there is scientific debate about their theory, and no evidence against it (Wells, 2000, p. 235f.)

Until the beginning of the 20th Century, evolution theory had no real chance of being taken seriously because of its laughably simple explanation. An unreasonable theory, such as that of macroevolution had no real chance of making a socio-political breakthrough – in the absence of demonstrable finds. This is why, then as now, scientific falsifications are required. After any number of years of falsification, a theory is now firmly anchored in people's minds, and is now so taken for granted that it is no longer questioned. Younger scientists are educated at our universities as theoreticians and trained experts like hamsters on a wheel. They learn certain laws, such as Haeckel's embryonic theory, by heart, but never question it because it is quite simply not allowed, if the student one day wants to graduate and then, perhaps, become a professor.

Because of the wealth of technical apparatus, measuring devices, probes and computers, the younger generation of scientists feels superior to the scientists of a hundred years ago who were connected, indeed rooted, in nature. Nonetheless, a specialised microbiologist today knows hardly anything about macroevolution and similar problematic subjects. Interdisciplinary training, which is necessary, is seldom to be seen. Connections are lost from sight and where the individual disciplines overlap, or even where this takes place within a single discipline, supposedly certain knowledge becomes warped.

It is almost exclusively interested outsiders who manage to draw attention to these dissonances and contradictions. These include scientists who research outside their given areas – interdisciplinary, so to speak. These scientists too are attacked by their colleagues until such time as they no longer dare to raise their heads above the parapet. Luckily, a few exceptions confirm the rule.

This is a systematic process: when the scandalous Old Stone Age falsifications at the *University of Frankfurt* were represented as the regrettable misconduct of a single, extremely fame–hungry scientist, the scientific community, particularly in cohort with the mass media, is giving itself carte blanche to continue scientific deceit, which is intentional and covered up by the mass media.

On 14 March 2005, the German television programme "nano" (*3sat* channel) it was said that a committee of investigation at the University of Frankfurt had uncovered Reiner Protsch's swindle, after years of hints from other scientists. This is also a lie, because there are no such monitoring mechanisms at universities. They are unwanted and would be counter–productive for the university system as a whole. In this connection the *Rheinische Merkur* newspaper (No. 37 of 9 Sep 2004) trenchantly observed that the university world and the possible perpetrators were to be protected, whilst the tardy "traitors" who revealed the swindle were to be attacked (cf. Illig, 2004, p. 499f.) – so the reverse of the "polluter-pays principle".

This is why falsifications which have been perpetrated over the last 30 years, with the knowledge of colleagues and co–authors who share the responsibility for them, have never caused any official stir. The committee of investigation at the University of Frankfurt was only called into being after the falsifications were published in Der Spiegel magazine, because the university is an independent, closed world, which protects its people. No action is taken against scientific swindles as long as third parties do not suffer financial damage. That is why Reiner Protsch was simply allowed to retire and everything just carries on as before.

Although the history of the Old Stone Age has apparently been invented from scratch, and the complete chronology and all of the theories based on it aren't worth the paper they're written on, it seems that this far–reaching falsification campaign is of secondary importance, indeed, trifling, to the *Gesellschaft der Anthropologie* ("Anthropology Society"). The Society's second Chairman, Prof. Carsten Niemitz (*Freie Universität Berlin*), even seems to find it necessary to say that the history of mankind should under no circumstances be rewritten. This is unadulterated ignorance and arrogance!

But it is not enough to hide a few fossilised Stone Age skulls in the cellar of the museum and cross out a few lines in the textbooks. The evolution lie makes it clear that the political codetermination and freedom of opinion in the sciences, for which people have been fighting since the Middle Ages, have gained absolutely no ground in terms of earth and human history (and other areas too?). This scientific ivory tower is in a similar position to that of the Roman Catholic Church before the case of Galileo. But this scientific construction of lies is full of gaping cracks and is going to collapse…

Bibliography

In addition to the quotations published in the journals (including *Science, Nature, PNAS, SpW, BdW*) the following authors are quoted:

Allchin, B., und Allchin, R.: "The Rise of Civilization in India Pakistan", Cambridge 1982

Ameghino, C.: "El femur de Miramar", in "Anales del Muso nacional de historia natural de Buenos Aires", 26, pp. 433–450, 1915

Arsuaga, J. L.: "El collar del neandertal", 1999, deutsch: "Der Schmuck der Neandertaler", Hamburg 2003

Austin, S. A.: "Grand Canyon: Monument to Catastrophe, Institute for Creation Research", Santee 1994, pp. 111–131

Austin, S. A.: "Excess Argon within Mineral Concentrates from the New Dacite Lava Dome at Mount St. Helens Volcano", in: "Creation Ex Nihilo Technical Journal", Vol. 10 (Part 3), 1996, pp. 335–343

Baales, M.: "Umwelt und Jagdökonomie der Ahrensburger Rentierjäger im Mittelgebirge", in: "Monographien des Römisch–Germanischen Zentralmuseums", 38, Mainz/Bonn 1996

Baales, M: "Economy and seasonality in the Ahrensburgian", 1999, in: Kozlowski, S. K. et al. (Publisher.): "Targed points cultures in Europe. Kolloquium Lublin", "Lubelskie materialy archeologiczne", 13, Lublin, pp. 64–75

Baigent, M.: "Das Rätsel der Sphinx", Munich 1998

Barbujani, G., und Bertorello, G.: "Genetics and the population history of Europe", in: "*PNAS*", Vol. 98, 2. 1. 2001, pp. 22–25

Bayer, J.: "Der Mensch im Eiszeitalter", Leipzig/Vienna 1927

Bayer, U.: "Die Erde unter Berlin ...", in: "Der belebte Planet", special publication of the FU Berlin, 2002, pp. 21–27

Becker, G. F.: "Antiquities from under Tuolomne Table Mountain in California", in: "Bulletin of the Geological Society of America", 2, 1891, pp. 189–200

Bent, W. Thomas: »The Tucson Artefacts«, USA 1964

Berg, A. von: "Ein Hominidenrest aus dem Wannenvulkan bei Ochtendung, Kreis Mayen–Koblenz", in: "Arch. Korrespondenzblatt", 27, 1997, pp. 531–538

Binford, L. R.: "Ancient Man and Modern Myths", New York 1981

Binford, S. R., und L. R.: "A preliminary analysis of functional variability in the Mousterian of Levallois faces", in: "American Anthropologist", Vol. 68, 1966, p. 238ff.

Bird, J.: "Antiquity and Migrations of the Early Inhabitants of South America", in: "The Geographical Review". Vol. 1, 6/1938, pp. 250–275

Blair, T. C.: "Alluvial–fan sedimentation from a glacial–outburst flood, Lone Pine, California, and contrasts with meteorological flood deposits", in: "Special Publication of International Association of Sedimentologists", Vol. 32, 2002, pp. 113–170

Bloess, C., und Niemitz, H.–U.: "C14–Crash", Gräfelfing 1997

Bloess, C.: "Ceno–Crash", Berlin 2000

Boll, E., und Brückner, G. A.: "Geogenie der deutschen Ostseeländer", Neubrandenburg 1846

Boman, E.: "Los vestigios de industria humana encontrados en Miramar (Republica Argentina) y atrbuidos a la época terciaria", in: "Revista Chilena de Historia y Geografia", Vol. 49 (34), 1921, pp. 330–352

Bord, J., und C.: "Unheimliche Phänomene des 20. Jahrhunderts", Rastatt 1995

Boule, M., und Vallois, H. V.: "Fossil Men", London 1957

Brace, C. L., Nelson, H., Korn, N., und Brace, M. L.: "Atlas of Human Evolution", New York 1979

Brandt, M.: "Ein Affe auf zwei Beinen: Oreopithecus", in: "Studium integrale Journal", March 1999, Year 6, Issue 1, pp. 33–37

Breuil, H.: "Sur la présence d'éolithes à la base de l'Éocène, Parisicii", in: "L'Anthropologie", Vol. 21, 1910, pp. 385–408

Brewster, D.: "Queries and Statements Concerning a Nail Found Imbedded in a Block of Sandstone Obtained from Kongoodie (Mylnfield) Quarry, North Britain", in: "Report of the Fourteenth Meeting of the British Association for the Advancement of Science", London 1845

Brock, B.: "Die Natur macht keine Sprünge? Das Denken schon!", in: Zillmer, H.–J.: "Mistake Earth Science", 2001, pp. 9–16

Brüning, H. "Die eiszeitliche Tierwelt von Mosbach, ihre Umwelt, ihre Zeit", "Museumsführer", 6, Mainz 1980

Bryan, A. L.: "Early Man in America from a Circum Pacific Perspective", Edmonton, Univ. of Alberta, Dept. of Anthropology, 1978

Bryant, E. A., et al.: "The impact of tsunami on the coastline of Jervis Bay", in: "Southeastern Australia. Physical Geography", Vol. 8, No. 5, 1997, pp. 441–460

Bryant, E. A., und Nott, J. A.: "Geological Indicators of Large Tsunami in Australia", in: "Natural Hazards", Vol. 24, No. 3, 2001, pp. 231–249

Büchner, M.: "Gesteinskundliches Gutachten: Grünsandstein. Fundort: Bison–Schlucht, Stonecreek, Kanada", in: Hoening, 1981, pp. 262–264

Bülow, K. von: "Abriss der Geologie von Mecklenburg", Berlin 1952

Burroughs, W. G.: "Human–like Footprints, 250 million years old", in: "The Berea Alumnus", Berea College (Kentucky), November 1938, p 46f.

Capellini, G.: "Les traces de l'homme pliocène en Toscana. Congrès International d'Anthropologie et d'Archéologie Préhistoriques", Budapest 1876, Vol. 1, 1877, pp. 46–62

Carling, P. A., et al: "Late Quaternary catastrophic flooding in the Altai Mountains of south–central Siberia: a synoptic overview and an introduction to flood deposit sedimentology", in: Martini, I. P. , et al.: "Flood and Megaflood Process and Deposits: Recent and Ancient Examples", Oxford 2002, pp. 17–35

Cavalli–Sforza, L. und F. "Verschieden und doch gleich", Munich 1994, TB 1996

Cavalli–Sforza, L. L.: "Gènes, peuples et langages", Paris 1996, deutsch: "Gene, Völker und Sprachen", Munich/Vienna 1999

Champion, T., et al.: "Prehistoric Europe", London 1984

Chard, C. S.: "An outline of the prehistory of Siberia", in: "Southwestern Journal of Anthropology", 14/1958, pp. 1–33

Charlesworth, E.: "Objects in the Red Crag of Suffolk", in: "Journal of the Royal Anthropological Institute of Great Britain and Ireland", Vol. 2, 1873, pp. 91–94

Credner, H.: "Elemente der Geologie", Leipzig 1912

Cremo, M. A., und Thompson, R. L.: "Forbidden Archaeology", Badger 1993; German: "Verbotene Archäologie", Augsburg 1997

Criswell, W. A.: "Stammt der Mensch vom Affen ab?", Wetzlar 1974, 2. 1976 Edition

Crompton, R., et al.: "The mechanical effectiveness of erect and 'bent–knee, bent–hip' bipedal walking in *Australopithecus afarensis*", in: "Journal of Human Evolution", Vol. 35, 1998, pp. 55–74

Czarnetzki, A.: "Neandertaler: Ein Lebensbild aus anthropologischer Sicht", in: "Neandertaler und Co. Ausstellungskatalog", Münster 1998

Dacqué, E.: "Die Erdzeitalter", Munich/Berlin 1930

Dalrymple, G. B.: "40Ar/36Ar analysis of historic lava flows. Earth and Planetary", in: "Science Letters", Vol. 6, 1969, pp. 47–55

Darwin, C.: "The Origin of Species", London 1859; German: "Die Abstammung der Arten", Licenced Edition Cologne 2000

Darwin, C.: "The Descent of Man", London 1871; German: "Die Abstammung des Menschen", Wiesbaden 1966

Dash, M.: "X–Phänomene", Munich/Essen 1997

Daubrée, G. A.: "Synthetische Studien zur Experimental–Geologie", Braunschweig 1880

Dawkins, R.: "The Blind Watchmaker", London 1986

Deloison, Y.: "Préhistoire du piéton – Essai sur les nouvelles origines de l'homme", Paris 2004

Deloria, V.: "Red Earth, White Lies", New York 1995

Derev'anko, A. P. , Shimkin, E. B., und Powers, W. R.: "The Paleolithic of Siberia: New Discoveries and Interpretations", Urbana 1998, pp. 336–351

Dillehay, T. D.: "The Settlement of the Americas", New York 2000

Dingus, L., und Rowe, T.: "The Mistaken Extinction – Dinosaur Evolution and the Origin of Birds", New York, 1997

Dorling Kindersley Ltd., Hrsg.: "Eyewitness Guide: Early People", London 1989, German: "Die ersten Menschen", Hildesheim 1994

Dubois, E.: "On the Fossil Human Skulls Recently Discovered in Java and Pithecanthropus Erectus", in: "Man", Vol. 37, January 1937

Dougherty, C. N.: "Valley of the Giants", Cleburne 1971

Egenolff, J. A.: "Historie der Teutschen Sprache", Leipzig 1735

Ehrenreich, P. : "Anthropologische Studien über die Bewohner Brasiliens", Braunschweig 1897

Engesser, B.: "Aufrecht auf zwei Beinen", in: "Basler Magazin – Politisch–kulturelle Wochenend–Beilage der Basler Zeitung", Basel, 8 August 1998, No. 30, pp. 1–5

Evernden, J. F., und Kistler, R. W.: "Chronology of emplacement of Mesozoic batholithic complexes in California and western Nevada", in: "U.S. geol. Surv. Prof. Paper, 1970", Vol. 623, p. 42ff.

Feder, K. L., und Park, M. A.: "Human Antiquity", Mountain View 1989

Fester, R.: "Die Eiszeit war ganz anders", Munich 1973

Fenton, S. R., Webb, R. H., et al.: "Cosmogenic 3He Ages and Geochemical Discrimination of Lava–Dam Outburst–Flood Deposits in Western Grand Canyon, Arizona", in: House, P. K., et al., 2002

Feustel, R.: "Abstammungsgeschichte des Menschen", Jena 1990

Flint, R. F.: »Glacial Geology and the Pleistocene Epoch«, New York 1947

Floss, H.: "Der Ziegenberg bei Altenrath. Ein Fundplatz der Ahrensburger Stielspitzengruppen am Südrand der Kölner Bucht", in: "Jahrbuch des Römisch–Germanischen Zentralmuseums Mainz", 34, 1989, p. 165–192

Fraas, E.: "Berichte über die Tendaguru–Expedition", in: "Sitzungsbericht der Gesellschaft naturforschender Freunde zu Berlin", 1909, No. 6, 358ff., No. 8, 500ff., No. 10, 631; 1910 No. 8, 372ff., 1912 No. 2b

Friedrich, H.: "Der Mythos von den angeblichen ›Rassen‹ der Menschheit", in: "EFODON Synesis", 6/1994, pp. 16–19

Gambier, D.: "Fossil hominids from the early Upper Palaeolithic (Aurignacien) of France", in: Mellar, P. , und Stringer, C. B.: "The Human Revolution", Edinburgh 1989, p. 194ff.

Gee, H.: "In Search of Time: Beyond the Fossil Record to a New History of Life", in: "The Free Press", New York, 1999, p. 126f.

Geise, G.: "Woher stammt der Mensch wirklich", Hohenpeißenberg 1997

Glover, I.: "Leang Burung 2. An Upper Palaeolithic Rockshelter in South Sulawesi, Indonesia", in:

"Modern Quaternary Research in Southeast Asia", Year 6, 1981, pp. 1–38

Gold, T.: "Das Jahrtausend des Methans", Dusseldorf/Vienna/New York 1988

Goren–Inbar, N.: "A figurine from Archeulian site of Berekhat Ram", in: "Mt–tequfat ha–even", Vol. 19, pp. 7–12

Gould, S. J.: "Men of the Thirty–Third Division", in: "Natural History", April, 1990, pp. 12–24

Gould, S. J.: "Illusion Fortschritt", Frankfurt 1998

Greely, A. W.: "Stefansson's Blond Eskimos", in: "National Geographic Magazine", Vol. XXIII, Vol. 12 December 1912

Greenberg, J. H., und Ruhlen, M.: "Der Sprachenstamm der Ureinwohner Amerikas", in: "Die Evolution der Sprachen", "Spektrum der Wissenschaft Dossier", 2004, pp. 58–64

Gregory, J. W.: "Contributions to the Geography of British East Africa", in: "Geographical Journal", IV, 1894

Gregory, J. W.: "The African Rift Valleys", in: "Geographical Journal", LVI, 1920

Gregory, R. L.: "Eye and Brain: The Physiology of Seeing", Oxford 1995

Gripp, K.: "Die Rengeweihstangen von Meiendorf", in: Rust, A.: "Das altsteinzeitliche Rentierjägerlager Meiendorf", Neumünster 1937, pp. 62–72

Grosswald, M. G.: "Late Weichselian Ice Sheet of Northern Eurasia", in: "Quaternary Research", 13/1980, p. 16

Haarmann, H.: "Universalgeschichte der Schrift", Frankfurt/New York 1991, Special Edition 1998

Haber, H.: "Unser blauer Planet", Stuttgart 1965

Hapgood, C. H.: "Maps of the ancient Sea Kings", Kempton 1966

Heinsohn, G.: "Wie alt ist das Menschengeschlecht?", München 1991, 4th Edition, 2003

Heinsohn, G.: "Für wie viele Jahre reicht das Grönlandeis?", in: "Vorzeit–Frühzeit–Gegenwart", Issue 4/1994, pp. 76–81

Heinsohn, G.: "Wann starben die Dinosaurier aus?, in: "Zeitensprünge", Issue 4/1995, pp. 371–382

Henke, R.: "Aufrecht aus den Bäumen", in: "Focus", 39/1996, p. 178

"Herder–Lexikon der Biologie", Heidelberg/Berlin/Oxford 1994, Vol. 6

Hilgenberg, O. C.: "Die Bruchstruktur der sialischen Erdkruste", Berlin 1949

Hilgenberg, O. C.: "Vom wachsenden Erdball", Berlin 1933

Hitching, F.: "The Neck of the Giraffe: Where Darwin Went Wrong", New York 1982, p. 204

Hoening, A. E. F.: "Fundort Stone Creek", Dusseldorf/Vienna 1981

Holmes, S. H.: "Phosphate Rocks of South Carolina and the Great Carolina Marl Bed", Charleston 1870

Holmes, W. H.: "Review of the evidence relating to auriferous gravel man in California", in: "Annual Report of the Board of Regents of the Smithsonian Institution for the Year Ending June 30, 1899", Part 1, Washington 1901, pp. 419–472

Homet, M. F.: "Die Söhne der Sonne", Freiburg 1958

Hooton, E.: "Apes, Men and Morons", New York 1937

House, P. K., Webb R. H., Baker V. R., und Levish, D. R.: "Ancient Floods Modern Hazards", Washington, DC 2002

Howell, C. F.: "Der Mensch der Vorzeit", Life – Wunder der Natur, 1969

Hrdlicka, A.: "Early Man in South America", Washington 1912

Hsü, K. J.: "Die letzten Jahre der Dinosaurier", Basel 1990

Hsü, K. J.: "Klima macht Geschichte", Zurich 2000

Hublin, J.–J., et al.: "A late Neanderthal associated with Upper Palaeolithic artefacts"; in: "*Nature*", 16. 5. 1996, pp. 224–226

Hublin, J.–J.: "Die Sonderevolution der Neandertaler", in: "Die Evolution des Menschen", "Spektrum der Wissenschaft Dossier", 2004, pp. 56–63

Illig, H.: "Die veraltete Vorzeit", Frankfurt 1988

Illig, H.: "C14: einmal mehr desavouiert. Causa Reiner Protsch von Zieten", in: "Zeitensprünge", 3/2004, pp. 497–502

Ivanov, A. A., Kouznetsov, D. A., und Miller, H. R.: Aus den Proceedings des 5. Internationalen Kreationisten–Kongresses in England, zitiert in: "Lebten Dinosaurier und Menschen zur selben Zeit?", in: "Factum", Issue 2/1993

Jacobi, A.: "Das Rentier, eine zoologische Monographie der Gattung Rengifer", in: "Zoologischer Anzeiger", Ergänzungsband 96, 1931

Jensen, H.: "Indogermanisch und Grönländisch", in: Hirt, H.: "Germanen und Indogermanen", Heidelberg 1936, p. 151ff.

Jia, L., und Huang, W.: "The story of Peking man", Beijing 1990

Johanson D. C., und Edey, M. A.: "Lucy: the beginnings of humankind", New York 1981

Johanson, D. C., Blake, E., und Brill, D.: "From Lucy to Language", 1996

Jordan, P. : "Die Expansion der Erde", Braunschweig 1966

Julig, P. J.: "The Sheguiandah Site: Archaeological, geological and paleobotanical studies at a Paleoindian site on Manitoulin Island, Ontario", in: "Mercury Series, Archaeological Survey of Canada", Paper 161. Hull 2002, Canadian Museum of Civilization

Junker, R.: "Stammt der Mensch vom Affen ab?", Stuttgart 1995, 7th Ed. 2002

Junker, R., und Scherer, S.: "Evolution: Ein kritisches Lehrbuch", Gießen 1998, 5th Ed. 2001

Kahlke, R.–D.: "Das Pleistozän von Untermaßfeld bei Meiningen (Thüringen)", Parts 1–3, Römisch–Germanisches Zentralmuseum Mainz 1997/2001

Keindl, J.: "Theorie der Weltraummassen", Vienna 1934

Keith, A.: "The Antiquity of Man", Vol. 1, Philadelphia 1928

Kelso, A. J.: "Physical Anthropology", New York 1970

Knußmann, R.: "Die mittelpalaeolithischen menschlichen Knochenfragmente von der Wildscheur bei Steeden (Oberlahnkreis)", in: "Nassauische Annalen", 68, 1967, pp. 1–25

Königswald, G. H. R. von: "Preliminary Report on a Newly–Discovered Stone Age Culture from Northern Luzon, Philippine Islands", in: "Asian Perspectives", II (2), 1956, pp. 69–71

Königswald, G. H. R. von: "Begegnungen mit dem Vormenschen", 1961

Kolosimo, P. : "Unbekanntes Universum", Wiesbaden/Munich 1991

Kutschera, U.: "Streitpunkt Evolution", Kassel 2004

Körber, H.: »Die Entwicklung des Maintals«, Würzburger geograph. Arb. 10, Würzburg 1962

Laing, S.: "Human Origins", London 1894

Leakey, M. D.: "Olduvai Gorge. Excavation in Beds I and II 1960–1963", Vol. 3, Cambridge 1971

LeBlond, P. H., und Bousfield, E. I.: "Cadborosaurus: Survivor from the Deep", Victoria 1995

Lee, T. E.: "The antiquity of the Sheguiandah site", in: "Anthropological Journal of Canada", 21, 1983

Lee, T. E.: "Untitled editorial note on the Sheguiandah site", in: "Anthropological Journal of Canada", 2(1), 1966, pp. 29ff. und 4(2), 1966, p. 50

Lembersky, M.: "Mount St. Helens", Portland 2000

Körber, H.: »Die Entwicklung des Maintals«, Würzburger geograph. Arb. 10, Würzburg 1962

Lindroth, C. H.: "The faunal connection between Europe and North America", New York/Stockholm 1957

Lippolt, H. J., et al.: "Excess argon and dating of Quaternary Eifel volcanism, IV: Common argon with high and lower–than atmosph. Ar40/Ar36 ratios", in: "Phonolitic rocks Earth.Planet.Sci.Letters", 101, 1, Amsterdam 1990, pp. 19–35

Lister, A.: "Mammoths", London 1994; deutsch: "Mammuts", Sigmaringen 1997

Lohest, M., et al.: "Les silex d'Ipswich", in: "Conclusions de l'enquête de l'Institut International d'Anthropologie", Vol. 33, 1923, pp. 44–47

Lyell, C.: "The Principles of Geology: Being an Attempt to explain the Former Changes of the Earth's Surface, by reference to Causes Now in Operation", London 1830, Reissued 1865

Lyell, C.: "Das Alter des Menschengeschlechts", Leipzig 1864

Macalister, R. A. S.: "Textbook of European Archaeology", Vol. 1, Cambridge 1921
Mania, D.: "Auf den Spuren des Urmenschen", Berlin 1990
Mason, J. A.: "The Ancient Civilizations of Peru", Harmondsworth 1957
Mayr, E.: "Einführung (zum Kapitel Makroevolution)", S. 319–322 in: Mayr, E.: "Eine neue
    Philosophie der Biologie", Munich 1991
Mijares, A. S.: "An Expedient Lithic Technology in Northern Luzon (Philippines)", in: "Lithic
    Technology", Vol. 26, No. 2, 2001, pp. 138–152
Moore, R.: "Die Evolution", in der Reihe: "Life–Wunder der Natur", 1970
Morgan, E.: "The Scars of Evolution", London, 1990
Mortillet, G. de: "Le Préhistorique", Paris 1883
Muck, O. H.: "Alles über Atlantis", Dusseldorf/Vienna 1976
Müller, W.: "Glauben und Denken der Sioux", Berlin 1970
Müller, W.: "Amerika, die neue oder die alte Welt", Berlin 1982
Müller–Beck, H.: "Urgeschichte der Menschheit, Stuttgart 1966
Müller–Beck, H.: "On migrations of hunters across the Bering Land Bridge in the Upper Pleistocene",
    in: Hopkins, D. M.: "The Bering Land Bridge", Stanford 1967, pp. 373–408

Niemitz, C.: "Das Geheimnis des aufrechten Gangs", München 2004
Nilsson, T.: "The Pleistocene", Dordrecht 1983
Noorbergen, R.: "Secrets of the Lost Races", New York 1977
Nummedal, A.: "Stone Age Finds in Finnmark", Oslo 1929
Oakley, K. P. : "Relative dating of the fossil hominids of Europe", in: "Bulletin of the British
    Museum" (Natural History), Geology Series, 34(J), 1980, pp. 1–63
Obermaier, H.: "Fossil Man in Spain", New Haven 1916, reissued 1924/1969
Oxnard, Ch. E.: "Uniqueness and Diversity in Human Evolution", 1975

Pachur, H.–J.: "Abschied von Eden", in: "Der belebte Planet", Sonderheft der FU Berlin, 2002,
    pp. 78–87
Paturi, F. R.: "Die Chronik der Erde", Augsburg 1996
Paturi, F. R.: "Die Chronik der Menschheit", Augsburg 1997
Pawlik, A. F.: "Is there an Earl Palaeolithic in the Philippines? New Approaches for Lithic", in:
    "Proceedings of the 7th Australasian Archaeometry Conference", 2001, pp. 255–270
Peiser, B. J.: "Was the Cambridge Conference a Flop? Evidence for Multiple Catastrophes in Historical
    Times", in: "Chronology and Catastrophism Review", Vol. 15, pp. 23–28
Pettijohn, F. J., Potter, P. E., und Siever, R.: "Sand and Sandstone", Berlin/Heidelberg/New York
    1972
Pitman, M.: "Adam and Evolution", London 1984
Pitman, W., und Ryan, W.: "Noah's Flood", New York 1998; deutsch: "Sintflut", Bergisch–Gladbach
    1999
Politis, G.: "Fishtail Projectile Points in the Southern Cone of South America", in: "Clovis: Origins and
    Adaptations", University of Maine, 1991
Potts, R., und Shipman, P. : "Cutmarks made by stone tools on bones from Olduvai Gorge,
    Tanzania", in: "*Nature*" Vol. 291, 18. 6. 1981, pp. 577–580
Probst, E.: "Deutschland in der Urzeit", Munich 1986, special edition 1999
Protsch, R.: "The Age and Stratigraphic Position of Olduvai Hominid I", in: "Journal of Human
    Evolution", Vol. 3, 1974, pp. 379–385

Ragazzoni, G.: "La collina di Castenedolo, sotto il rapporto antropologico, geologico ed agronomico",
    in: "Commentari dell'Ateneo di Brescia", 4. April 1880, pp. 120–128

Raup, D.: "Conflicts Between Darwin and Palaeontology", in: "Bulletin, Field Museum of Natural History", Vol. 50, January 1979, p. 24

Renfrew, C.: "Die Indoeuropäer – aus archäologischer Sicht", in: "Die Evolution der Sprachen", "Spektrum der Wissenschaft Dossier", 2004, pp. 40–48

Renfrew, C.: "Die Sprachenvielfalt der Welt", in: "Spektrum der Wissenschaft", Juli 1955, p. 72ff.

Reynolds, S. J., et al.: "Compilation of Radiometric Age Determinations in Arizona", in: "Arizona Bureau of Geology and Mineral Technology Bulletin", 197, 1986

Ridley, F.: "Transatlantic Contacts of Primitive Man. Eastern Canada and Northwestern Russia", in: "Pennsylvania Archaeologist", 1960

Ries, G.: "Schlechte Kronzeugen", Internet: Newsgroup: de.sci.geschichte, 18. 5. 2003, 20:51:54 GMT

Ripota, P. : "Was Charles Darwin uns alles verheimlichte", in: "P. M. Magazin", Issue 04/2002

Ritters, V.: "Ein halb geschälter und versteinerter Seeigel", in: "EFODON Synesis", 30/1998, pp. 7–11

Romero, A. A.: "El Homo Pampaeus", in "Anales de la Sociedad Cientifica Argentina", 85, pp. 5–48, 1918

Ronquillo, W. P. : "The Technological and Functional Analysis of Lithic Flake Tools from Rabel Cave, Northern Luzon, Philippines", in: "Anthropological Papers", No. 13, National Museum Manila 1981

Roth, S., et al.: "Acta de los hechos más importantes del descubrimento de objetos, instrumentos y armas de piedra, realizado en las barrancas de la Costa de Miramar, partido de General Alvarado, provincia de Buenos Aires", in "Anales del Museo de historia natural de Buenos Aires", 26, pp. 417–431, 1915

Rust, A.: "Das altsteinzeitliche Rentierjaegerlager Meiendorf", Neumünster 1937

Rust, A.: "Vor 20000 Jahren. Rentierjaeger der Eiszeit", Neumünster 1962

Rutte, E.: »Die Fundstelle altpleistozäner Säugetiere von Randersacker bei Würzburg«, in: »Geol. Jb. 73«, 1958, S. 737–754

Rutte, E.: »Die Fossilstellen des Mittelmaincromer im stratigraphischen Vergleich mit den benachbarten Fundstellen«, in: »Quartärpaläontologie 8«, Berlin 1990, S. 233–236

Sanford, J. T.: "Sheguiandah reviewed", in: "Anthropological Journal of Canada", 9(1), 1971, pp. 2–15

Sarre, F. de: "The Theory of Initial Bipedalism – On the question of Human origins", "Biology Forum", Rivista di Biologia, Università di Perugia, 87 (2/3), pp. 237–258, Perugia 1994

Saurat, D.: "L'Atlantide et le regne des géants", Paris 1955/1969

Schlosser, M.: "Beiträge zur Kenntnis der oligozänen Landsäugetiere aus dem Fayum", in: "Beiträge zur Paläontologie und Geologie", Vol. 24, pp. 51–167

Schmidt, E.: "Vorgeschichte Nordamerikas im Gebiet der Vereinigten Staaten", Braunschweig 1894

Schoetensack, O.: "Der Unterkiefer des Homo Heidelbergensis aus den Sanden von Mauer bei Heidelberg", Leipzig 1908

Schwarzbach, M.: "Das Klima der Vorzeit", Stuttgart 1993

Sepharim, E. Th.: "Das Pleistozänprofil der Kiesgrube Kater in Hiddesen bei Detmold. Ein prä– moränales Schotterkonglomerat mit Gletscherschliff", in "21. Ber. naturwiss. Verein Bielefeld", Bielefeld 1973, pp. 249–263

Sergi, G.: "L'uomo terziario in Lombardia", in: "Archivio per l'Antropologia e la Etnologia", 14, 1884, pp. 304–318

Shapiro, R.: "Origins: A Sceptics Guide to the Creation of Life on Earth", New York 1986

Shin, Y., et al.: "Synthesis of SiC Ceramics by the Carbothermal Reduction of Mineralized Wood with Silica", in: "Advanced Materials", Vol. 17, Issue. 1, January 2005, pp. 73–77

Simonsen, P. : "The Rock Art of Arctic Norway", in: "Bolletino del Centro Camuno di Studi Preistorici", 11/1974, pp. 129–150

Simpson, G. G., und Beck, W.: "An Introduction to Biology", New York, 1965

Sinclair, W. J.: "Recent investigations bearing on the question of the occurrence of Neocene man in the auriferous gravels of the Sierra Nevada", in: "University of California Publications in American

Archaeology and Ethnology", 7(2), 1908, pp. 107–131

Spears, I. R., und Crompton, R. H.: "The Mechanical Significance of the Occlusal Geometry of Great Ape Molars in Food Breakdown", in: "Journal of Human Evolution", Vol. 31, 1996, pp. 517–535

Spetner, L. "Continuing an exchange with Dr. Edward E. Max", http://www.trueorigin.org/spetner2.asp, 2001

Spriestersbach, J.: Lenneschiefer (Stratigraphie, Fazies und Fauna). – Abhandlungen des Reichsamts für Bodenforschung, Neue Folge, 203, Berlin 1942

Stan, H., und Hess, J. C.: "Physical foundations of dating by the K–Ar– and Ar40/Ar39 methods", in: "Course book of Isotope Geology", Breslau 1990, pp. 184–98

Standen, A.: "Science is a sacred cow", New York 1950

Stansfield, W. D.: "The Science of Evolution, New York 1977

Stefansson, V.: "My life with the Eskimo", London 1913

Stolyhwo, K.: "Le crâne de Nowosiolka considéré comme preuve de l'existence à l'époque historique de formes apparentées à H. Primigenius", in: "Bulletin International de l'Académie des Sciences de Cracovie", 1908, pp. 103–26

Strauss, S.: "Systems give boost to dating technology, in: "The Globe and Mail", Toronto, 2. 4. 1991, A12

Stringer, C. B., und Gamble, C.: "In search of the Neanderthals: Solving the Puzzle of Human Origins", New York 1993

Tattersall, I.: "The Fossil Trail", New York/Oxford 1995

Tattersall, I.: "The Last Neanderthal", New York 1995. 2nd Ed. 1999

Taylor, G. R.: "The Great Evolution Mystery", New York 1983

Taylor, J.: "Fossil, Facts and Fantasies", Crosbyton 1999

Thenius, E.: "Die Evolution der Säugetiere", Stuttgart 1979

Thomas, A.: "Les secrets de l'Atlantide", Paris 1969

Thompson, K. S.: "Ontogeny and Phylogeny Recapitulated", in: "American Scientist", Vol. 76, May/June 1988, p. 273

Thorne, A. G., und Macumber, P. G: "Discoveries of Late Pleistocene man at Kow Swamp", in: "*Nature*", Vol. 238, 1972, pp. 316–319

Trinkaus, E., und Shipman, P. : "The Neanderthals: Changing the Images of Mankind", New York 1992; German: "Der Neandertaler. Spiegel der Menschheit", Gütersloh 1992

Trinkaus, E.: "Hard Times Among the Neanderthals", in "Natural History", Vol. 87, December 1978

Trinkaus, E: "The Mousterian Legacy: Human Biocultural Change in the Upper Pleistocene", in: "British Archaeological Reports International", Vol. 164, 1983, pp. 165–200

Turner, E., et al.: "Neandertaler oder Höhlenbär? Eine Neubewertung der ›menschlichen‹ Schädelreste aus der Wildscheur", in: "Hessen. Archäologisches Korrespondenzblatt", 30, 2000, pp. 1–14

Tuttle, R. H.: "Natural History", Maerz 1990

Velikovsky, I.: "Earth in Upheaval", Garden City 1955; German: "Erde im Aufruhr", Frankfurt/M. 1980

Vening Meinesz, P. A.: "Spanningen in de aardkorst door poolverschuivingen", in: "Afd. Natuurk.", Vol. 52, 1943, pp. 186–196

Vogel, K.: "The Expansion of the Earth – An Alternative Model to the Plate Tectonics Theory", in: "Critical Aspects of the Plate Tectonics Theory"; Vol. II, "Alternative Theories", Athens 1990, pp. 14–34

Vogl, D.: "Der Darwinfaktor", Greiz 2001

Wahnschaffe, F.: "Die Ursachen der Oberflächengestaltung des norddeutschen Flachlandes", 1891, 4th Ed. edited by Ms. Schucht. 1921

Walther, J. W.: "Geschichte der Erde und des Lebens", Leipzig 1908

Weaver, T. D.: "The shape of the Neanderthal femur is primarily the consequence of a hyperpolar body form", in: "*PNAS*", Vol. 100, 10 June 2003, pp. 6926–6929

Wells, J.: "Icons of Evolution: Science or Myth? Why Much of What We Teach About Evolution is Wrong", Regnery Publishing, 2000

Wendt, H.: "From Ape to Adam", Indianapolis 1972

Williams, Stephen: »Fantastic Archaeology. The wild side of North America Prehistory«, Philadelphia 1991/2001

Willis, E.: "Fossils and Phosphate Specimens", 1881

Whitney, J. D.: "The auriferous gravels of the Sierra Nevada of California", in: "Museum of Comparative Zoology Memoir", 6(1), Harvard University 1880

Wong, K.: "Der Streit um die Neandertaler", in: "Die Evolution des Menschen", "Spektrum der Wissenschaft Dossier", 2004, pp. 64–71

Wright, F.: "Man and the Glacial Period", New York 1897

Yahya, H.: "Der Evolutionsschwindel", Istanbul 2002

Young, R. W., Bryant, E. A., und Price, D. M.: "Catastrophic wave (tsunami?) transport of boulders in southern New South Wales, Australia", in: "Zeitschrift für Geomorphologie", Vol. 40, No. 2, 1996, pp. 191–207

Zarate, M. A., und Fasana, J. I.: "The Plio–Pleistocene record of the central eastern Pampas, Buenos Aires, Province, Argentina: the Chapadmalal Case Study", in: "Paleogeography, Paleoclimatology, Paleoecology", 7/2 1989, pp. 27–52

Zilhao, J., und d'Errico, F.: "Die unterschätzten Neandertaler", in: "Die Evolution des Menschen", "Spektrum der Wissenschaft Dossier", 2004, pp. 68–69

Zillmer, H.–J.: "*Darwin's Mistake*", Enkhuizen 2002, 2nd Ed. 2003, original German title: "Darwins Irrtum", Munich 1998, 9th Ed. 2009

Zillmer, H.–J.: "Mistake Earth Science", Victoria 2007, original German title: "Irrtümer der Erdgeschichte", Munich 2001, 5th Ed. 2008

Zillmer, H.–J.: "Dinosaurier Handbuch" (Dinosaur Handbook: not yet published in English), Munich 2002

Zillmer, H.–J.: "The Columbus-Mistake", in progress to be publish in English, German title: "Kolumbus kam als Letzter", Munich 2004, 3rd Ed. 2009

Zillmer, H.–J.: "Der Energie-Irrtum" (The Energy Mistake), Munich 2009

Züchner, C.: "Archäologische Datierung. Eine antiquierte Methode zur Altersbestimmung von Felsbildern?", in: "Quartär", 51/52, 2001, pp.   107–114 und Vortrag auf der 42. Tagung der Hugo–Obermaier–Gesellschaft für Erforschung des Eiszeitalters und der Steinzeit e.V., Tübingen, 25–29 April 2000

Zuckerman, S.: "Beyond The Ivory Tower", New York 1970

Index

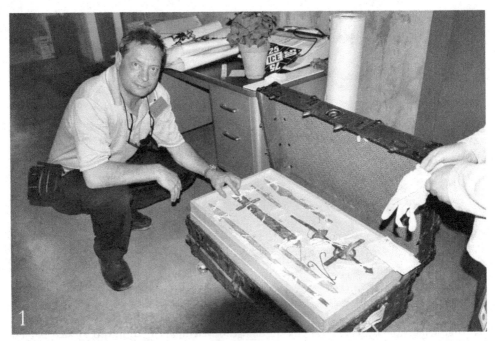

1 Hans-Joachim Zillmer opens the crate containing the Silverbell artefacts in the museum cellar of the Arizona Historical Society. On one of the swords excavated in 1924 there is a depiction of a sauropod in "modern" horizontal posture, a stance which has only been recognised by science in the last few years.

2 The individual layers containing various artefacts are lifted.

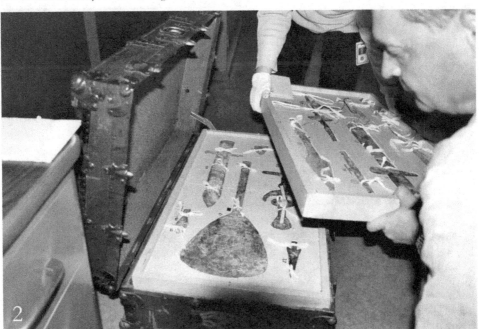

3 A large number of clay sculptures depicting dinosaurs were found in Acambaro, Mexico. In these depictions, people are feeding dinosaurs like house pets. Several sculptures were examined in various laboratories. Dating showed them to be at least 2,000 years old. Official reconstructions of dinosaurs, however, are less than 200 years old.

4 Michael Cremo (author of "Forbidden Archaeology") and Hans-Joachim Zillmer (right hand side) after their lectures at the "Ancient Mysteries" exhibition in Vienna. Cremo documented finds of human remains from geological strata going back 600 million years.

Human cadaver bone at 2,000X.

Specimen EC96-001 at 2,000X.

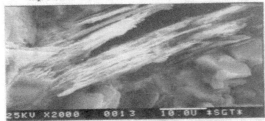

While examining anthracite boulders above ground close to Mahanoy City and Shenandoah (Pennsylvania), Ed Conrad discovered a large boulder which appeared to contain a human skull. The Smithsonian Institution ruled that it was not a human relic and that the find was not even a bone. It was "established" that the find was a concretion. The range and type of examination carried out were not disclosed. However, enlarging the "concretion" by 2,000 times reveals that it is very similar in structure to human bones. The authenticity of this find would contradict evolution theory, as the "skull" is in rock which is over 280 million years old. Conrad also found dinosaur relics in the same geological layers and deduces from this that man and dinosaur must have co-existed - although there were supposedly no dinosaurs 280 million years ago.

5 Human bones found by Ed Conrad in the anthracite boulder.

6 Enlargement of a normal, human bone structure.

7 Enlargement of the Conrad skull concretion.

8 The "concretion" (left) in comparison with a normal human skull as can be found in old stone strata (right).

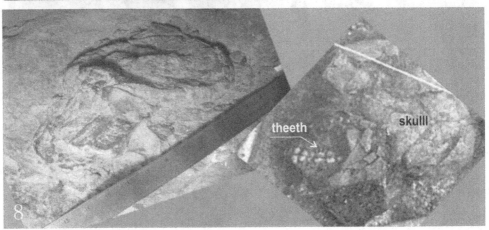

theeth

skulll

9 The impression of a human hand was discovered in a 140 million year old Chalk Age sandstone layer at the Paluxy River in Glen Rose, where dinosaur footprints are also in evidence.

10 Carl E. Baugh (left hand side), the Director of the Creation Evidence Museum and Hans-Joachim Zillmer in front of the dinosaur-era hammer, discovered in Chalk Age sandstone near London, Texas, which was described in detail in "Darwin's Mistake". It consists of 96.6% pure steel (tempered steel) without carbon and the wooden shaft is fossilised.

11, 12 In 2000, on removal of a rock layer, the layer beneath revealed a line of footprints from dinosaurs and three people, whereby one of the human footprints was in the middle of a three-toed impression (size comparison with Larissa Zillmer's foot). Since the best print at Taylor Trail was destroyed, the print shown here was removed and the original can now be seen at the Creation Evidence Museum (Glen Rose). Klaue = Claw of dinosaur foot.

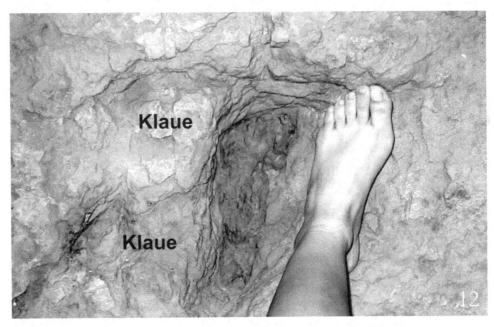

13

13 At the beginning of 1969 a huge head was washed up in Tecolutla (Mexico). It was approximately four metres high, weighed one ton and was thought to be the head of a sea snake.

14 A 19th Century etching shows the sighting of a sea snake at Cape Ann in Massachusetts in 1639. It was the first sighting of a sea snake in the New World. There were many subsequent sightings of aquatic monsters with similar descriptions.

14

15 A copy of an old rock drawing is on display at Maine State Museum in Augusta, Maine. American Indian legends in many States speak of a terrifying monster which lives in the water. (cf. Fig. 9).

A Monstrous Sea Serpent,

The largest ever seen in America,

Has just made its appearance in Gloucester Harbour,

Cape Ann, and has been seen by hundreds of

Respectable Citizens.

The Editor of the Salem Gazette, says:—We have in our possession an extract of a letter from John Low, Esq. to his son in this town, dated Gloucester, Thursday afternoon, August 14, 1817.

"There was seen on Monday and on Tuesday morning playing about our harbor, between Eastern Point and Ten pound Island, a SNAKE with his head and body about eight feet out of water, his head is in perfect shape as large as the head of a horse, his body is judged to be about FORTY-FIVE OR FIFTY FEET IN LENGTH, it is thought he will girt about 3 feet round the body, and his sting is about 4 feet in length.

While writing the above a person has called in, who says that there are two to be seen, playing from the Stage-head into the harbor inside of Ten pound Island.

The spectators are Mr. Charles Smith, Mr. John Proctor and several others. A number of our sharp shooters are in pursuit of him but cannot make a ball penetrate his head, Another party is just going in pursuit with guns, harpoons &c. Our small craft is fearful of venturing out a fishing.

The above can be attested to by twenty different people of undoubted veracity."

In addition to this account the Salem Register states, that the Serpent is extremely rapid in its motions which are in all directions, that it shews a length of 50 feet; that a man who discharged his musket within 30 feet of the Serpent, says its head was partly white and that he bit it, that a large sum had been offered for it ; that "it appears in joints like wooden buoys on a net rope almost as large as a barrel, that musket balls appear to have no effect on it, that it appears like a string of gallon kegs 100 feet long."

The editor of the Register quotes an account of a Sea Serpent seen on the coast in 1746, something like it. It had a head like that of a horse, and as he moved he looked like a row of large casks following in a right line.

The Boston Daily Advertiser in speaking of this *Monstrous Serpent*, says:—We have seen several letters from Gloucester, which describe a prodigious Snake that has made its appearance in Cape-Ann Harbour. It was first seen by some fishermen, 10 or 12 days ago, but it was then generally believed to be the creature of the imagination. But he has since come within the harbor of Gloucester, and has been seen by hundreds of people. He is declared by some persons who approached within 10 or 15 yards of him, to be 60 or 70 feet in length, round, and of the diameter of a barrel. Others state his length variously, from 50 to 100 feet. His motions are serpentine, extremely varied, and exceedingly rapid. He turns himself completely round almost instantaneously. He sometimes darts forward, with his head out of water, at the rate of a mile in 3 minutes, leaving a wake behind, of half a mile in length. His head, as large as the head of a horse, is shaped somewhat like that of a large dog, is raised about 8 feet out of water and is partly white, the other part black. He appears to be full of joints and resembles a string of buoys on a net rope, as is set in the water to catch herring. Others describe him as like a string of water casks. His back is black. Various attempts have been made, without success, to take him. Four boats went out on Thursday, filled with adventurous sailors and experienced gunners, armed with muskets, harpoons, &c. Three muskets were discharged at him from a distance of 30 feet, two balls were thought to hit his head, but without effect. He immediately after plunged into the water, and disappeared for a short time, after which he moved off to the outer harbor, and was seen no more that night. A number of persons are employed in making a net of cod-line, of sufficient size to take him. It is conjectured that he has resorted to this harbor for the purpose of preying upon a very numerous shoal of herrings, which have lately appeared there. If he has been instrumental, as is supposed, in driving these herring into the harbor, he has rendered an essential service to the town.

The Salem Gazette of the 19th inst. says, " We are informed, that on Sunday this creature was seen playing sometimes within 15 or 20 feet of the shore, affording a better opportunity to observe him than had before occurred. Gentlemen from Gloucester state, that he appeared to them even of greater magnitude than had before been represented, and should judge from their own observation, that he was as much as 100 feet in length, and as big round as a barrel. They saw him open an enormous mouth, and are of opinion that he is cased in shell.

Aug. 22, 1817.

Printed and Sold by Henry Bowen, Devonshire-Street, Boston.

16 On 22 August 1817 there were reports of a sea snake of around 20 metres in length, which had been seen by many people several times on various days. Boats were organised to undertake a search (see picture), so that the sea snake could be shot at close range. The body was said to be as round as a barrel with a circumference of 90 centimetres. The head, the size of a horse's head, was said to look like a dog. Similar reports of a sea snake had already been made in 1746.

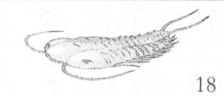

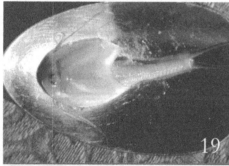

17 The author visits Mabel Meister, now deceased. On splitting open a 570 million year old slate, her husband discovered a shoe print. A trilobite was visible on the sole. These crustaceans died out 250 million years ago with the start of the dinosaur era.

18 Drawing of a type of trilobite.

19 In "Mistake Earth Science" (Zillmer, 2007) the question was asked as to whether this presently-living animal is a trilobite. It is, however, a Triops longicaudatus. This crustacean is closely related to the trilobite. They have existed for at least 220 million years without any kind of evolutionary development right up to the present day. If stones contain trilobites they should be hundreds of millions of years old. Isn't it possible that like triops, trilobites still existed until recently? Did a man wearing shoes step on the "Master trilobite" only a few thousand years ago? Then the slate has to be relatively young - otherwise it means that people were already wearing shoes at the time of the "Cambrian Explosion" 570 million years ago.

20 This stone from the dinosaur era contains typical human extremities and was found near Bogotà, Colombia.

ancient
Hopi Lake

ancient seafloor

Hopi Lake

C

dinosaur prints

C

ancient seafloor

23

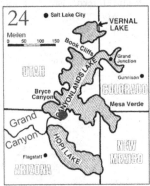

24 The pre-Flood lakes in the Grand Canyon were drained at the time of the dinosaurs. Near Tuba City (black circle), at the edge of Lake Hopi, footprints, coproliths (C) and dinosaur bones lie in what is today desert.

21–23 Allegedly, dinosaurs lived in the shallow shore waters of Lake Hopi 140 million years ago. They left thousands of footprints in the thin chalk layer which at that time, as shown by the fossilised ripple markings, was the floor of the lake. Countless fossilised piles of dung (coproliths) are scattered around. How long does it take before a previously soft pile of dung fossilises without any sign of erosion?

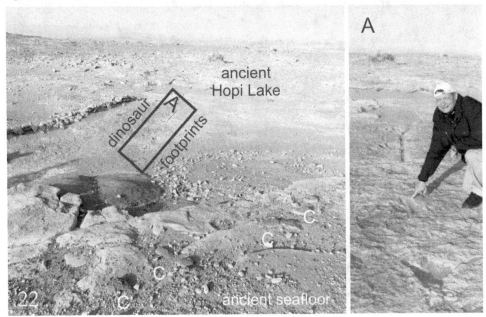

ancient
Hopi Lake

A

dinosaur footprints

C

C

C

C

ancient seafloor

22

A

25 The "International Colloquium on the Earth's Expansion - Examining a Theory" was held from 24-25 May 2003 at the Mining and Industry Museum at Theuern Castle in Eastern Bavaria. Photo: Zillmer (2003).

26 A globe of an expanding earth; one of a series of glass models created by Klaus Vogel from Werdau, which shows how the oceans grew continuously, starting from a closed continental earth crust which enclosed a small earth landmass.

27 A stone quarry near Bellambi(Australia) shows an old mud deposit, only a few thousand years old, from the Holocene Period, as proof of a superflood. This deposit is covered by a sandy layer from the Pleistocene Period. The third layer comprises a more recent sand layer from the Holocene Period. Photo: Ted Bryant.

28 Aerial view of the castle wall of Arkona (Rügen)from the north-east. The ring wall of what was once a much bigger construction, broke down with the collapse of the chalk cliffs. Photo: National

29, 30 When Mount St. Helens in the State of Washington erupted, huge masses of mud rolled down into the valley and created layers of up to 50 metres deep in which tree trunks were trapped (oval). Water which then came out of the volcano dug deep "canyons" in these layers, so that table mountains were created. Insert above: View of the volcano and flow direction of mud and water (arrow).

31 Deep canyons with broken off edges and steep cliff faces were created in only a few hours by fast erosion - similar to the Grand Canyon.

32 Through the explosion, the tree trunks were snapped like matchsticks and carried off by the waves of mud. Insert: cars being carried off by the mud torrent, like the dinosaurs were earlier.

33

33 The lower jaw of Heidelberg Man was found in 1907 at the Rösch sandpit near Mauer on Elsenz. An entire humanoid-type with untold generations was constructed on the basis of this fragment. The white cross marks the spot where the jawbone was discovered. However, these Cromer layers were "filled in, in a single filling phase" (Rutte, 1990, p. 235), "without interruption in a very short time geologically speaking (Körber, 1962, p. 30). This may have occurred only a few thousand years ago.

34 In 1933, the Steinheiner Woman's skull was found in the Sigrist sandpit in Steinheim (marked with a cross). The skull was said to be 250,000 years old. However, it was found in a layer of gravel deposited by the Murr, in which elephants were also discovered. All of the layers from the earth's surface down to the skull were gravel layers from the river - was this going on for

35 Example of a time impact in 1980. This field of rubble was created within 15 minutes after the Mount St. Helens eruption at a speed of 65 kilometres per hour. This type of rubble with rounded debris is usually proof of an Ice Age. It was not the little stream visible in the foreground which created the rubble instantaneously, but gigantic floodwaters.

36 Aerial photo of the Sierra Nevada (in background), California, in which many controversial, anomalously old artefacts have been found in the table mountains. Mount Whitney (W) is 4,417 metres in height. Superfloods washed huge masses of mud from the Hogback Canyon (H) and Tuttle Canyon (T) in to the Sierra Piedmont Plain, around the Alabama Hills (foreground) (arrows) and through the valleys right up to the town of Lone Pine (L). Photo: Edited from Blair, 2002.

37 The Grand Canyon was cut out of the plateau by large masses of water in several phases. This occurred only a short time ago, in geological terms, most recently 1,300 years ago. The edges of the Grand Canyon are sharp and steep.

38 The old flat plain of the Rhinian Slate Mountains, "recently cut" (Dacqué, 1930) by large masses of water from the Rhine.

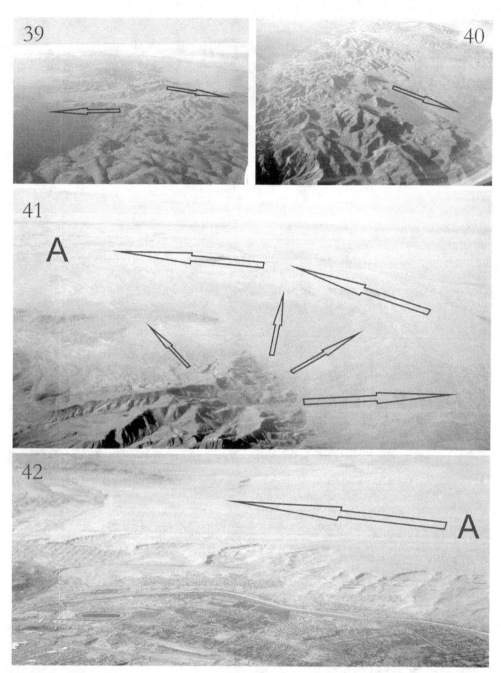

39–42 During his flights across western America, the author took photographs of the landscape. Mud flow, which came from the hills and mountains down to the high plains and made the landscape into an infertile waste, can be seen clearly. The mud layers show almost no erosion. It therefore appears that this event took place relatively recently within a short space of time. Where do these massive quantities of material, which must previously have been on the mountains, come from? Did the mountains burst through the previously flat layers only a few thousand years ago?

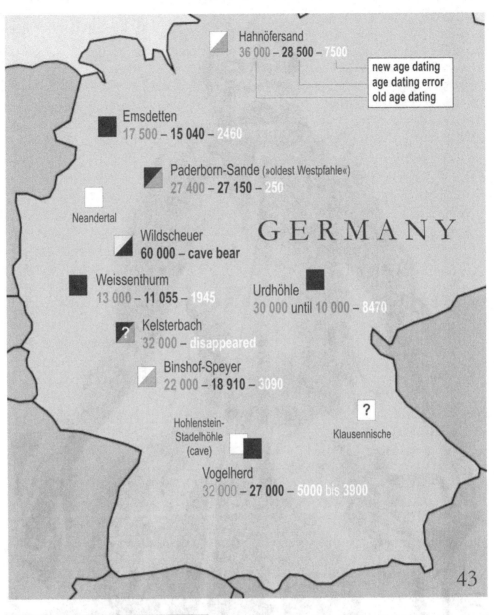

Hahnöfersand
36 000 – **28 500** – 7500

new age dating
age dating error
old age dating

Emsdetten
17 500 – **15 040** – 2460

Paderborn-Sande (»oldest Westpfahle«)
27 400 – **27 150** – 250

Neandertal

GERMANY

Wildscheuer
60 000 – cave bear

Weissenthurm
13 000 – **11 055** – 1945

Urdhöhle
30 000 until 10 000 – 8470

Kelsterbach
32 000 – **disappeared**

Binshof-Speyer
22 000 – **18 910** – 3090

Klausennische
?

Hohlenstein-
Stadelhöhle
(cave)

Vogelherd
32 000 – **27 000** – 5000 bis 3900

43

Legend:

- modern human: new age dating (C-14)
  old age dating on the basis of stratigraphy
- modern human: new age dating (C-14)
  old age dating on basis C-14 dating
- Formerly classified as Neanderthals
  however, misinterpretation
- Neanderthals: new age dating (C-14)
  old age radiocarbon dating (C-14)
- ? Neanderhals: questionable Fund (tooth)
- Leftover bones of Neanderthals
  (not yet re-examined)

43 It is said that modern Man arrived in Central Europe 35,000 years ago. New investigations show, however, that as predicted in "Darwin's Mistake", most of the bones from the more recent Stone Age (including those of Neanderthal Man) are considerably younger and belong to the New Stone Age (Note: Middle Stone Age = Phantom Age). There may be older bones but these are rarities, from which one cannot extrapolate a history of "pre-Flood" Man. Amended illustration based upon "Der Spiegel" (34/2004).

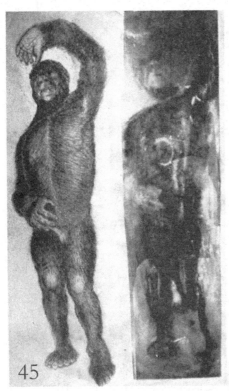

44

45 This 1.8m tall "Ice Man" was allegedly fished out of the Bering Sea on the east coast of Siberia in a massive block of ice. According to US Marine press reports, however, the giant ape was shot in Vietnam, then flown to America and frozen. The American journalist, Ivan T. Sanderson, photographed it and it was examined by the zoologist, Bernard Heuvelmans. Later, the body was replaced by an imitation and it was shown at fairs in America. Did the original body have to disappear because as a "living yeti" he didn't fit into the history of Man's evolution? Photo left: reconstruction. Right: original photo.

46 Example of a very large Acheulean hand-axe in eastern Morocco (Saurat, 1955, Table 10). Does the discovery of such finds infer that there were big people in the past?

44 Left-hand side: in Venezuela, an expedition was attacked by two giant apes. One of them was shot and was posed sitting, propped up with a stick, to be photographed. The giant ape was 1.6m tall and had 32 teeth. In 1996, "The Anomalist" magazine tried to represent these mysterious creatures as mere spider monkeys. But for this the animals are too tall and the teeth are an anomaly. Moreover, there are no apes in America, because the settlement of the continent allegedly began only a few thousand years ago and the absence of apes in America is seen as the proof that the evolution of man took place in Africa.

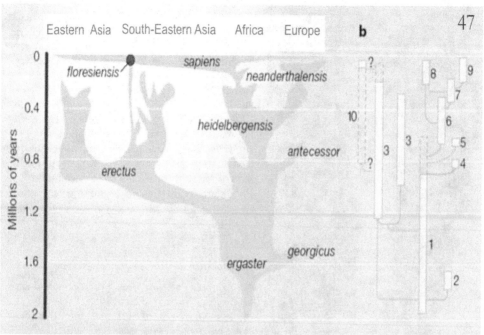

Eastern Asia   South-Eastern Asia   Africa   Europe   **b**   47

floresiensis   _sapiens_   _neanderthalensis_   8 9 7

heidelbergensis   10   6   7

antecessor   3   3   5

erectus   ?   4

Millions of years: 0, 0.4, 0.8, 1.2, 1.6, 2

georgicus   1

ergaster   2

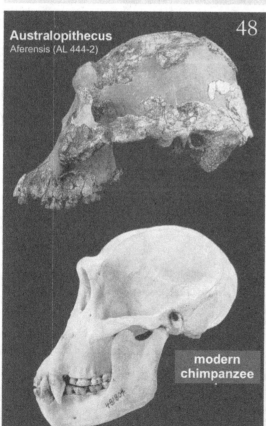

Australopithecus
Aferensis (AL 444-2)   48

modern chimpanzee

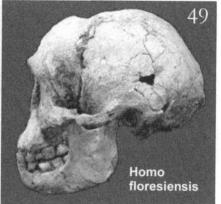

49

Homo floresiensis

47 The development of the human species according to (Lahr/Foley in "Science News", 27.10.2004): Homo ergaster/Africanus erectus; 2. Homo georgicus; 3. Japanese and Chinese Homo erectus; 4. Homo antecessor; 5. Home cepranensis; 6. Homo heidelbergensis; 7. Homo helmei; 8. Homo neanderthalensis; 9. Homo sapiens; 10. Homo floresiensis.

48 The skull of Australopithecus afarensis (above) does not differ from that of a modern chimpanzee.
49 Skull of Homo floresiensis.

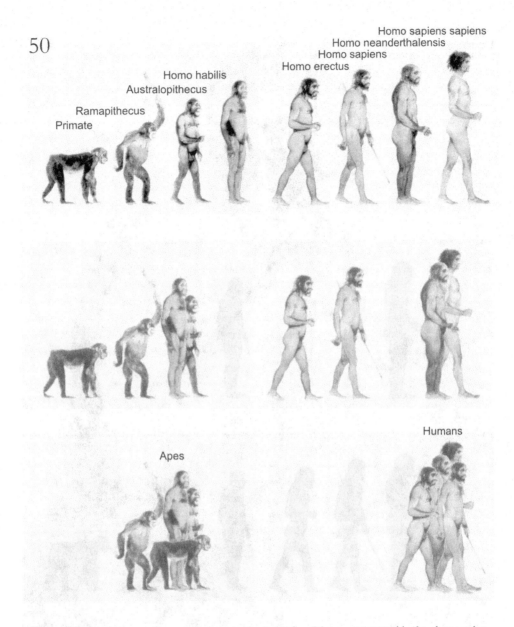

50 The evolutionary stages from common ancestor to modern Man are presented in the above order, showing the dogma of evolution theory as it existed until 1997. According to latest research, Homo habilis has been reclassified as an ape (Australopithecus habilis). Neanderthal man was ruled out as Man's ancestor and is classified as a distinct species (Homo neanderthalensis) or as a sub-species of modern Man (Homo sapiens neanderthalensis). The lower row shows the opinion expressed in this book, that there are only humans and apes. All human species, from Homo erectus to modern Man are only different variations of modern Man through adaptation and micro-evolution. Sketch from "Chronik der Menschheit" (1997).

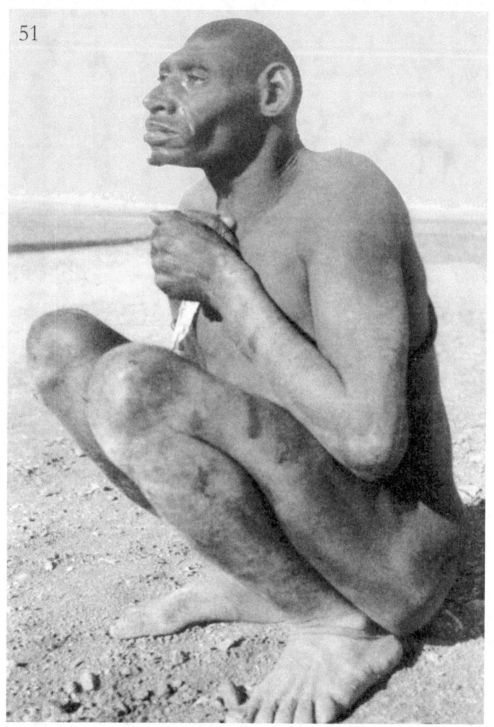

51 In 1931, a "prehistoric" human of Neanderthal type with receding forehead, receding chin and very prominent browridge. He went naked, used only rudimentary tools, lived in a cave and ate raw meat. Photo: Marcel Homet, 1958.

AGE B.P.	CULTURE	PERIOD	EXAMPLE
12 000	Late Magdalenian V-VI	Classic (Stil IV)	
	Middle Magdalenian III-IVI		
17 000	Early Magdalenian I-II	Archaic (Stil III)	
	Solutrean		
	transition period	Primitive (Stil II)	
22 000			
27 000	Gravettian	Primitive (Stil I)	
32 000	Aurignacian		

52 It is said that the development of artistic representations of horse's heads during the more recent Old Stone Age took around 20,000 years according to current dating (new after Leroi-Gourhan in "Scientific American", 1968, Vol. 218, No. 2, p. 63). Gunnar Heinsohn (2003, p. 87) asked whether 1,000 rather than 20,000 years would not have sufficed for this development. Is this figure too high as well? Wouldn't only a few generations, at most 200-500 years, be adequate?

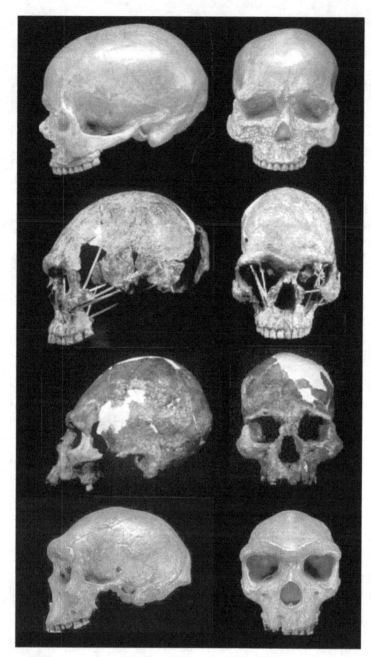

The skull found in 1868 in Les Eyzies (Dordogne) is classified as early-modern (Cro-Magnon 1), and until recently was said to be 30,000 years old.

Two skulls found in the Kow Swamps in Australia, Kow Swamp 1 and Kow Swamp V.

The so-called Rhodesian Man (Broken Hill 1) was discovered in 1921. Today he is classified as

53 Two skulls were discovered on 10 October 1967 in the Kow swamps in Victoria (Australia). The finders, Alan Thorne and Phillip Macumber classified both as Homo sapiens' skulls, although they showed many similarities with Homo erectus (see comparison with a Homo heidelbergensis skull). There were, however, no similarities with early-modern Man (see Cro-Magnon skull). The sole reason why they were regarded as Homo sapiens is because according to calculations they were only around 10,000 years old. The evolutionists did not want to accept the fact that Homo erectus, who for them was a primitive forerunner model, lived side by side with modern Man for years, until recently.

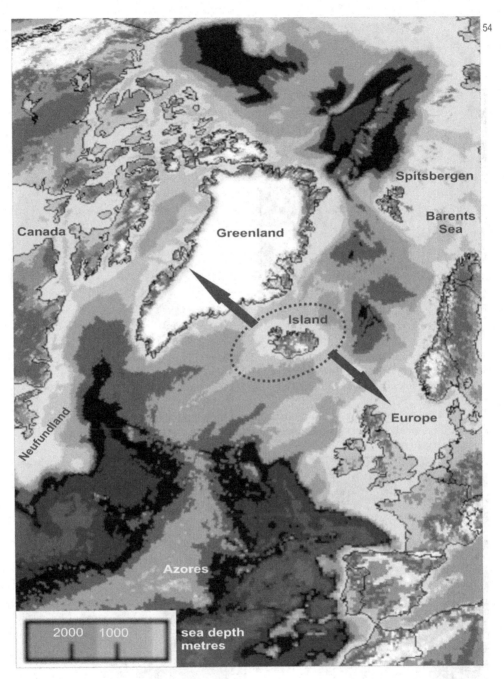

54 The depth chart of the North Pole sea around Iceland illustrates the sinking of the entire basin north of the Atlantic, discovered by Fridjof Nansen. Today only the highest points are above sea level. The adjacent floes sank isostatically at the same time and today are once again moving upwards, including Newfoundland. Before this happened, America and Europe were connected by the Greenland Bridge. At that time, the North Sea, the Baltic, the continental shelf areas west of England and Ireland and the Barents Sea were settled. Chart: NOAA NGDC, 15.11.1999.

239

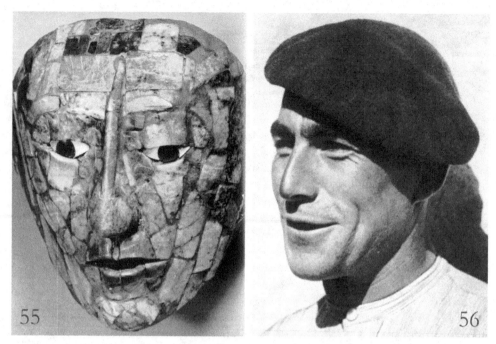

55 | 56

55 The jade mask from the king's tomb at the "Temple of Inscriptions" in Palenque (Mexico).
56 Basques could be the great-grandchildren of the "Man in the Jade Mask": look at the bold sweep of the typically aquiline nose, the expression of the eyes and the shape of the mouth. From Muck, 1978, p.201.
57 "Two Guns White Calf", a Blackfeet Indian Chief with typically-shaped nose.
58 A member of the Aymara people, the original inhabitants of the Bolivian highlands, with typically-shaped nose.

57 | 58

59 The cap-wearing stone idols of Pokotia belong to the first cultural period of Tiahuanaco(Bolivia).

59

60

60 In 1915, John Layard photographed an aborigine beside a huge menhir on the South Sea island of Malekula in Vanuatu, north-east of Australia.

61 Giant sculptures on Easter Island with narrow faces and prominent browlines, long noses and long ears. Insert: the girl's head (left) was reconstructed from an ivory carving (found at Unerwisternitz, Mähren in Germany). The prominent browline, long nose, small mouth, strong chin and longish head are particularly interesting. On Easter Island (right) one finds strange ancestral heads of a similar type. The Easter Islandwriting resembles the pictograms at Mohenjo-Daro, Indus Valley. Are these worldwide reflectors of a global culture?

61

62 Map of the distribution of kaftans (Oriental outer clothing) and ponchos (cape with head hole)in northern Eurasia and North America according to Müller (1982). It may be seen that the use of the poncho did not spread from Europe via Siberia to America. Did it originate in the previously warmer Arctic and move in a north-south direction to Europe and America?

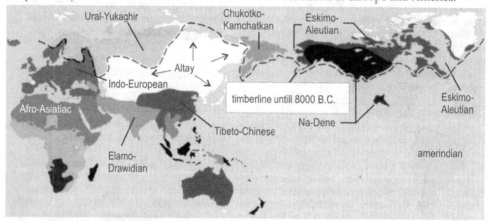

63 The spread of Indo-European (dark blue) related languages(green) in two separate climate-dependent phases. The first phase took place in a north-south direction (Na-Dené) starting from the Arctic and the second was made by the Eskimos in an east-west direction into eastern Siberia. (after Renfrew, 2004, p.29).

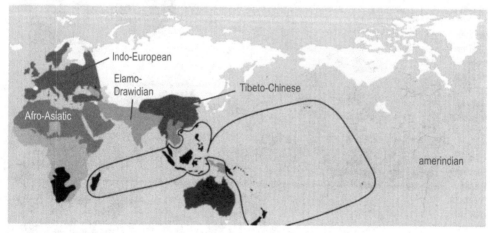

64 The languages of the land tillers spread. Larger families developed, inter alia Indo-European, Elamo-Dravidian, Tibeto-Chinese and Afro-Asiatic. (After Renfrew, 2004, p.29).

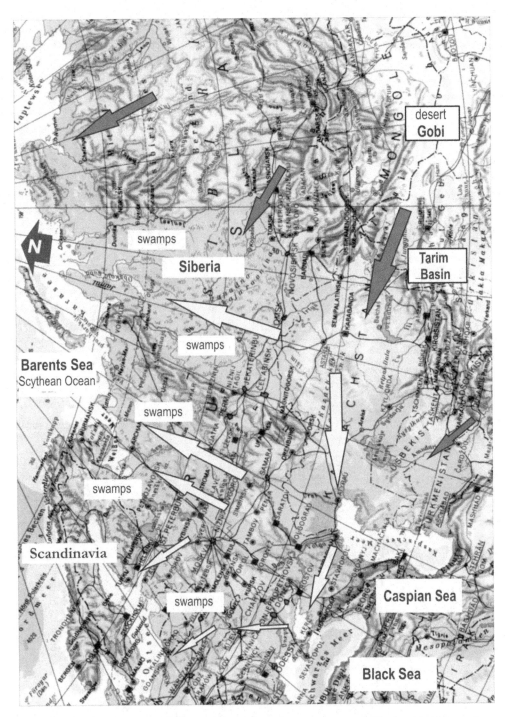

65 Mountain superfloods (dark grey arrows) are shown to have occurred in the Altai Mountains in south-central Siberia (Carling et al., 2002). The light gray arrows show the drainage of the water in the direction of the Barents Sea and the Black Sea. The Caspian Sea is connected to the "Scythian Ocean" (Barents Sea) through the Caspian Depression.

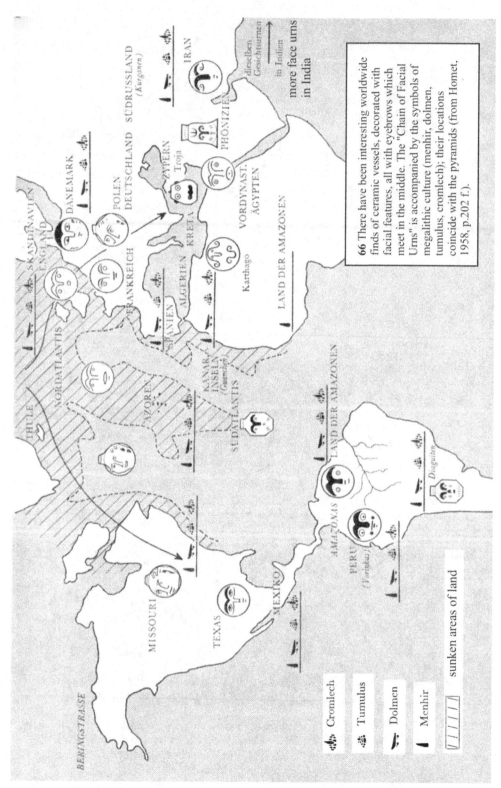

66 There have been interesting worldwide finds of ceramic vessels, decorated with facial features, all with eyebrows which meet in the middle. The "Chain of Facial Urns" is accompanied by the symbols of megalithic culture (menhir, dolmen, tumulus, cromlech); their locations coincide with the pyramids (from Homet, 1958, p.202 f.).

Cromlech
Tumulus
Dolmen
Menhir
sunken areas of land

more face urns in India

BERINGSTRASSE

THULE

MISSOURI

TEXAS

MEXIKO

NORDATLANTIS

AZOREN

SÜDATLANTIS

KANAR. INSELN (Guanchen)

SKANDINAVIEN

ENGLAND

DANEMARK

POLEN

DEUTSCHLAND

SÜDRUSSLAND (Kurganen)

FRANKREICH

ZYPERN

KRETA

Troja

PHÖNIZIEN

IRAN

SPANIEN

ALGERIEN

Karthago

VORDYNAST. ÄGYPTEN

LAND DER AMAZONEN

AMAZONAS

LAND DER AMAZONEN

PERU (Vorinka)

Diaguiten

244

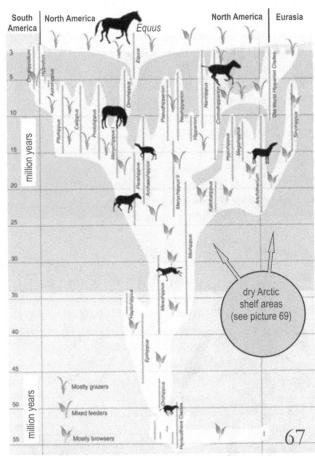

| South America | North America | | | North America | Eurasia |

Equus

million years

3
5
10
15
20
25
30
35
40
45
50
55

million years

dry Arctic
shelf areas
(see picture 69)

Mostly grazers

Mixed feeders

Mostly browsers

67

67 The current family tree of the horse shows the evolution of the horse which developed in North America. Horse species repeatedly spread to the Old World via the Greenland Bridge and the Bering Straits. Whilst the horses died out in the Old World, one species of horse (Equus) survived in America and split into two separate lines three million years ago (inter alia zebras, wild horses)and "these spread as far as the Old World" ("Science", Vol. 307, 18.3.2005, pp.1728-1730). According to Edgar Dacqué (1930, p.515), however, specialised species of horse originate from the previously warm, Arctic regions.

68 In the region of today's Sahara Desert, many rock drawings depict herds of cattle, elephant, rhinoceros and ostrich. The photo shows a detail from a 3m long cave painting, which is estimated to be 6,000-7,000 years old. At that time, the Sahara Desert was still fertile land.

68

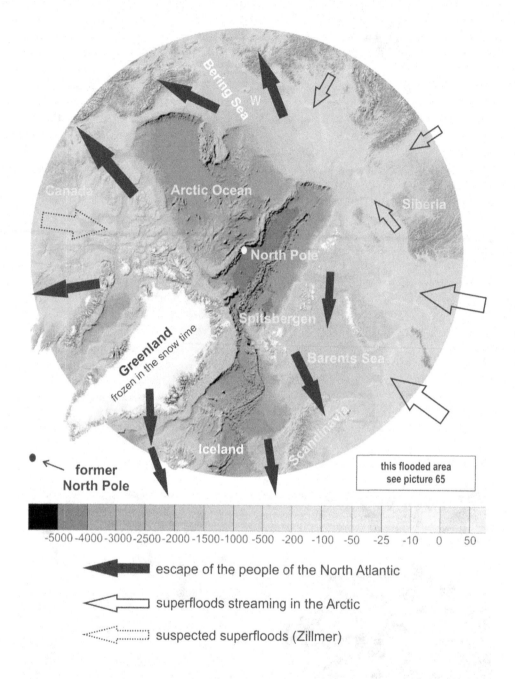

Legend within figure:

Bering Sea
W
Arctic Ocean
Canada
Siberia
North Pole
Spitsbergen
Greenland
frozen in the snow time
Barents Sea
Iceland
Scandinavia

this flooded area
see picture 65

• ← former
North Pole

-5000 -4000 -3000 -2500 -2000 -1500 -1000 -500 -200 -100 -50 -25 -10 0 50

→ escape of the people of the North Atlantic

⇨ superfloods streaming in the Arctic

⇢ suspected superfloods (Zillmer)

69 The North Pole previously lay further south from Greenland. Today's Arctic regions were free of ice. Wide areas in the north were settled. Since the sea level rose, only a few areas of land rise above the water as islands, such as Spitzbergen. Today's continental shelf areas were made uninhabitable by the encroaching cold and the subsequent flooding. The settlers fled to Europe, East and West Siberia, Beringia and North America. During the snow period, Greenland froze up quickly, as proven by the flora conserved under the ice. Mammoths continued to survive until 3,700 years ago on Wrangel Island (W) which is today within the central summer pack ice border. Topographical chart: NOAA NGDC, 15.2.2005.

A German Bestseller: **Mistake Earth Science**.

In his bestseller, *Darwin's Mistake*, which has meanwhile been translated into eleven languages, the independent private scholar and consulting civil engineer Dr. Hans-Joachim Zillmer has proved that there were megafloods processes and cataclysms but no macroevolution, whereas *Mistake Earth Science* presents a global revolution in the truest sense of the word in an exciting format. It's a lot of fireworks of facts and evidence, but also "live" excavations with the author and visits to all continents permit a radical revision of previously imagined worlds in favor of newer, trend-setting models of thought.

The various alternative models of thought that were previously introduced by the author have meanwhile been scientifically confirmed: a shift in the earth's axis of at least 20 degrees at the time of the dinosaurs, and also some cataclysms - the "global" deluge? - at the time of human life, because genetic investigations have shown that mankind had once almost died out. Or also, for example, that birds simply don't descend from dinosaurs. Here, the author also documents each piece of evidence that is to be checked out against the theory of evolution, which thereby ends as a beautifully fictional fairy-tale.

In this book, geological and geophysical scenarios and facts are introduced that are still widely relatively unknown. How does one explain that what was previously regarded as the "cradle" of evolutionary theory is geologically too young? Because the Galapagos Islands are only a few million years old and do not stem from the time of the dinosaurs as Charles Darwin erroneously suspected. Why were the largest hippos ever swimming in the rivers of Central European, and this was during the great ice age as well? Dinosaurs once lived on all continents, even on Spitsbergen and in Alaska and at the South pole, since there was a tropical climate from pole to pole. And now even today, the polar cap ice is melting down dramatically rapidly. In a few thousand years, there will not be any more ice, just as it was during the Mesozoic era, at both poles. More and more new finds of the same dinosaur species on almost all continents, but also on both sides of the Atlantic, have put continental drift in the form in which it has been promoted until now into question. On the basis of the most recent NASA research, there could have been geo-electrical events on earth that were previously not considered possible: is continental drift an incorrectly interpreted event? Oceans virtually devoid of water, a Mediterranean that dried up, even the presence of hippos on islands such as Cyprus and other phenomena are scenarios that are not to be explained by our world view and currently observable events. The incredible claim advanced by the best-selling author of this book, that the Amazon formerly originated out of the Sahara and fed into the Pacific, was already confirmed scientifically while the book was in press. Whoever is interested in the origins and development of our planet, as well as our present biosphere, won't be able to extract himself from the fascination of this logical demonstration of evidence and will see the earth's history with completely new eyes.

272 pages; quality trade paperback (softcover); catalogue #07-0105, Trafford Publishing
ISBN 1-4251-1735-X; **US$ 34.90**, C$ 40.14, EUR 27.21, £ 18.04

Coming in spring 2010:
A German bestselling book of

HANS-JOACHIM ZILLMER

THE COLUMBUS-MISTAKE

At the time when Greenland was green:
Old-Europeans in America

In his new book Hans-Joachim Zillmer presents astonishing findings and amazing insights. Celts and Vikings populated Greenland when it was still green. Normans imported Brazil wood. Christians travelled to South America and through the Magellan Strait into the Pacific Ocean long before Columbus. Do world-wide megaliths, signaling towers and the network of broad causeways, which connect the Mayan cities, testify of common roots? Why were the Inca rulers often light-skinned with red or blond hairs?

During his elaborate travels Zillmer made sensational discoveries. Those interested in the mysteries of human history won't be able to withdraw from the fascination of Zillmer's patterns of interpretation.

THIS BOOK SHAKES THE PILLARS OF OUR
HISTORIC KNOWLEDGE.
Prof. Dr. Wolfgang Kundt, University Bonn in Germany